THE ESKIMO

AND THE

OIL MAN

OTHER NONFICTION BY BOB REISS

*The Coming Storm: Extreme Weather and
Our Terrifying Future*

The Road to Extrema

*Frequent Flyer: One Plane, One Passenger, and the
Spectacular Feat of Commercial Flight*

THE ESKIMO

AND THE

OIL MAN

THE BATTLE AT THE TOP OF THE WORLD
FOR AMERICA'S FUTURE

BOB REISS

**BUSINESS
PLUS**

NEW YORK   BOSTON

Business Plus
Hachette Book Group
237 Park Avenue
New York, NY 10017

www.HachetteBookGroup.com

Printed in the United States of America

RRD-C

First Edition: May 2012

10 9 8 7 6 5 4 3 2 1

Business Plus is an imprint of Grand Central Publishing.
The Business Plus name and logo are trademarks of Hachette Book Group, Inc.

The publisher is not responsible for websites (or their content) that are not owned by the publisher.

Library of Congress Cataloging-in-Publication Data

Reiss, Bob.
    The Eskimo and the oil man : the battle at the top of the world for America's future / by Bob Reiss. — 1st ed.
        p. cm.
    ISBN 978-1-4555-2524-9
    1. Itta, Edward.   2. Shell Oil Company.   3. Petroleum industry and trade—Alaska.   4. Petroleum reserves—Alaska.   5. Eskimos—Land tenure—Alaska.   6. Offshore oil well drilling—Alaska.   I. Title.

HD9567.A4R45 2012
333.8'230916327—dc23
                                                                            2011052365

For Noah Lev Schoolsky,
a great guy with a great future

CONTENTS

THE ESKIMO

AND THE

OIL MAN

Chapter 1

The Argument, April 2010

The stakes could not have been higher.

Late on the morning of April 29, 2010, as the sun rose above the Arctic Ocean, a worried 64-year-old Iñupiat Eskimo whale hunter named Edward Itta stepped from a tent set into solid ice 320 miles above the Arctic Circle. He stood five feet above the Chukchi Sea. Behind him the ice stretched four miles to land. Ahead lay a channel of black water separating him from more pack ice further out. It led 1,200 miles later to the North Pole. His crew of eight had been camped here for days waiting to hunt bowhead whales migrating west from northern Canada, past Alaska.

"I did not want to leave," he would later recall.

Itta reluctantly climbed aboard his blue Arctic Cat snowmobile and steered the two-stroke machine roughly south along the meandering trail that Eskimo crews had hewn into the ice by hand. The backbreaking labor with picks and axes had taken weeks. The four-foot-wide path was marked by small red plastic flags and Itta's destination was Barrow, Alaska, a city of 4,500, northernmost municipality in the United States, where he had been born and where he had to attend a private meeting that day with three executives from one of the richest corporations on earth. He feared that their plan—if it went wrong—could destroy his 4,000-year-old community. Yet he feared that he might have to support it.

"I had been losing sleep over this for months."

Itta, five foot ten, wore a white pullover whaling Windbreaker over a thick lambskin parka, goggles to protect his face against the ten-degree temperature and frostbite, thermal underwear, and extra-warm Northern Outfitters boots. His feet were susceptible to cold due to an old injury, a broken ankle. The sun would stay up for two hours in late April. Since the end of January, when it reappeared over Barrow after months of winter dark, it had remained visible a few more minutes each day. The fresh light was pinkish and welcome. At times it could seem almost like a tropical veneer over a dusting of fresh snow.

"My father first took me out to a whaling camp when I was seven years old. I couldn't go out in the boat then. I ran errands and helped."

Arctic ice had been thicker then, Iñupiats knew, had formed earlier in the fall each year and melted later each spring. Temperatures had been colder. Back in the 1950s, Iñupiats had used dogs to haul the wooden sleds that carried supplies to hunters and chunks of whale meat back to feed the town. But by 2010 snowmobiles were the norm and several were parked behind Itta's crew's tent, hidden from the sea so the bright colors wouldn't scare off whales.

"The ocean is our garden." It supplied the fish, seals, whales and walrus that kept the Iñupiats alive.

Itta loved whaling; the camaraderie, the physical labor, the life at camp. Dinner last night had been duck soup and boiled seal meat cooked over a portable propane stove. Twenty-four hours a day the men took turns watching for whales or sleeping on warm animal skins in their clothing, ready to move instantly if a bowhead showed up. The massive mammals had passed for centuries each spring, coming from winter feeding grounds, heading for Canada from the Bering Strait.

"Our elders tell us to take the younger ones. They are plumper and tastier."

If a suitable bowhead came close—Barrow's 39 crews were scattered along the ice edge—the men would launch their eighteen-foot *umiaqs* or traditional open boats. They would use wooden paddles to approach the animal. The boats had no motors. They were usually made of sealskin stretched over wooden frames. The harpooner would be in front, the gunwale bobbing inches from the sea. A mature bowhead can reach 70 feet in length and each step of the hunt was perilous.

One reason Itta wished he could stay at camp was that he was captain and a captain's job was to keep his men safe. His crew were mostly family, including his 28-year-old son, Price. Itta had purchased their equipment and food, personally loaded explosive powder into their foot-long, missile-shaped, whale-killing projectiles. He'd picked the camp's location.

Each of these choices represented a decision on safety. Choose the wrong spot and the ice could break off and float away with the crew on it. Load the powder wrong and the explosive dart would be a dud; the whale could survive the initial attack and turn on the boat or drag it out to sea. It could capsize the boat or crush the crew. Even dead, a whale could cause an accident while the tricky work of hauling it back to camp was accomplished. Lines could get tangled. Men could be dragged down.

Iñupiats had been hunting whales in northern Alaska for thousands of years and every man knew the stories. They had been passed on by elders and were recorded in dances and recounted at parties and potluck gatherings that went on year round in the eight villages of the North Slope.

The men on their way to see Itta had never been on a whale hunt. They would arrive in a private jet.

The landscape was gorgeous. The ice world through which Itta passed showed the solid geometry of rock. There were rubble fields and ice-slick ravines. The treads sank slightly in mushy areas and slid metallically when crossing bare spots. The trail meandered through a magnificent rough region of jumbled ice boulders, massive blocks thrusting skyward where two ice fields had collided. The resulting pressure ridge had formed in the same way that mountains are created when one belt of rock hits another. The ice was an extension of the continent. It appeared each year with the predictive regularity of an eons-old natural clock.

And the colors! At times a deep blue seemed to shine out from beneath the earth. The undersides of ice boulders were stained brown with algae. The sky went gray-white and a fine mist of snow began blowing. A dim yellow light ahead meant another snowmobile was coming, and then passing, its bundled driver waving a mitten as he steered by.

Grinding through four-foot-high drifts—thinking about what he would say in the meeting—Edward carried a shotgun over his shoulder for protection against polar bears. They prowled the ice in search of food, seals usually, but during whaling season a hauled-up carcass represented tons of fresh meat. Polar bears tend to shy away from noisy snowmobiles but are capable of hunting a human. Other crews had spotted one the day before near Itta's camp.

Still, bears were a daily fact of life in Barrow. Itta was less concerned about them and more with the peril posed by the trio coming from Anchorage and Houston, North American headquarters of Shell Oil. The men had not explained exactly why they had asked for this meeting. "They never do. They said they were coming just to talk. But they'll want something."

The bottom line today was that Itta *loved* whaling and he'd rather be outside than meeting oil men in an office. But when he'd confided his desire to his old boyhood pal and crewmember Bart Ahsogeak last night, Bart had replied, "You have to go. You're the one everybody trusts. You'll know the right words. You're the voice of the people."

Now Edward was paying more attention to his problem than to where he was going.

This would soon cause an accident.

He reached shore and the landscape flattened out and he accelerated across ice-covered beach, ice-sheathed lakes and softer tundra. The snow was not deep except for piled drifts. Technically Barrow's environs are considered desert, and little precipitation falls each year. What lands doesn't melt. Winds move it around. Snow is so dry it clings to stop signs and rooftops in town like sand after a windstorm in Arizona. Plows nightly move piles of snow back to where they had been the day before. During whiteouts, snowmobile operators navigate by using snowdrift angles as a compass. Wind is a finger pointing toward home, or away.

Barrow is not reachable by road, just by tundra, sea or air. Coming up on Wiley Post Airport, which receives two Alaska Airlines jets each day, Edward saw the lone runway and beyond it, one- and two-story wooden homes of the city. There were light poles and gravel roads. Two multi-story office buildings, the Wells Fargo Bank and the headquarters of the Eskimo-owned Arctic Slope Regional Corporation, which operated businesses across the United States, dominated the view.

Suddenly Edward felt himself losing control. The snowcat crashed down on him, pinning him.

"I thought, You dumb shit. Pay attention."

He managed to climb out and right the Cat. Snowmobiles on Barrow streets are a common sight, and he made it back to his small wood-sided house. His leg ached from the fall but he'd suffered no injury. Barrow homes are perched on blocks to keep them from melting the permafrost below. Inside they are heated by natural gas from a local field, and cozy when winter temperatures drop as low as 70 below zero. Boots stay in the foyer. Parkas get hung on hooks.

Itta's residence was comfortable, furnished with stuffed chairs and a clutter of electronics in the living room: TV, computer, exercise walker. A floral-motif comforter covered a couch. The paneled walls—as in most Barrow homes—were a photo gallery of family members.

Off came the whaling gear as a transformation began. The hunter was changing into a political leader who met on occasions with Alaskan senators, Navy admirals, foreign diplomats, the secretary of the interior and even White House officials. Itta was morphing into the elected mayor of Alaska's North Slope Borough, a Wyoming-sized county populated by 7,500 people, mostly Iñupiats. He turned on the TV as he dressed in an open-necked, button-down blue shirt and jeans. Without the parka he looked slighter, with thick salt-and-pepper hair parted in the middle, wire-rimmed glasses and a flesh-colored hearing aid in his left ear.

What he saw on screen made his heart freeze up. There was an oil platform in a tropical sea, but then the platform was in flames, sinking, and he saw an armada of boats trying to stop spilled oil from spreading on the water. The burning rig was called the *Deepwater Horizon*.

What if that happens here, he thought?

Itta guessed that today's meeting would concern the explosion. Around the world hundreds of millions of viewers were watching the same scenes in the Gulf of Mexico.

But most viewers did not have Itta's responsibility to a community on their shoulders. Itta envisioned the whaling crews camped north of Barrow. The whole city was waiting for the first catch of spring.

At the Presbyterian church Itta attended each Sunday, sermons included pleas like, "We need more whales." At the high school, the basketball team was the Whalers. The city's Iñupiat Heritage Center was a museum featuring whaling artifacts and even photos of Edward, years earlier, throwing a harpoon at a whale.

Whales were the base of his community, and now the vivid oil spill on TV merged with concerns that had been torturing Itta for months.

This was because the men coming to see him planned to sink one exploratory well in summer 2010, in waters off the North Slope. If that went smoothly, Shell hoped to drill one or two more. Shell people had told Itta that they believed oil lay as close as eighty miles from where he had been hunting whales. The company had obtained most of the necessary federal permits to begin the exploration. They had leased a drillship that was currently in the Philippines but would soon sail north.

"If we lose the ocean, we have lost the Iñupiat Eskimo. A spill would wipe out thousands of years of our culture."

Itta worried about more than just a spill. He worried that Shell's seismic work might drive sound-sensitive whales away. That pollution from ships might cause respiratory illness in North Slope coastal villages. That helicopters transporting workers once operations began might scare off caribou, seals and whales. Food-wise, that would be the equivalent of an outside force emptying every supermarket and convenience store in any US city.

As Itta left his house, his concerns were shared by most residents of the region. Subsistence hunting was not only the basis of Iñupiat culture but it provided the food that people ate. In the last census 61 percent of residents who worked full time and 89 percent of the unemployed reported getting over half their nourishment from hunting and fishing.

What to do? Fight Shell or not? The whale hunter—in his

political capacity—was one of the most influential rural mayors in the United States. At his orders borough lawyers had challenged Shell in court in 2007, charging the federal agency responsible for permitting any offshore drilling with failing to conduct underlying science, failing to show whether the drill plan would do harm to the areas offshore.

"Too much, too fast, too soon," Itta had said then, and the court had agreed with him.

A Shell spokesman said that year, "That we failed I lay directly at the feet of Edward Itta."

But now Shell had changed the plan—made it smaller, and promised to stay away longer during hunting weeks—so Itta had refused to join national environmental groups—and a few Eskimo ones—still trying to bar Shell in court. His problem was *not* that he wished to halt all oil development. It was far trickier. It was a microcosm of energy issues facing the world.

"We need some oil development," Itta said.

By 2010, oil companies contributed over $250 million in taxes to the borough annually, from onshore facilities operating at Prudhoe Bay. This money paid for the airport he'd passed that day, and schools, plumbing, home heating and road clearance in winter throughout the region. Oil revenue comprised almost the entire North Slope budget.

Itta favored onshore development, but onshore fields were drying up in Alaska. The easy-to-get stuff was gone. If the decline continued the revenue would stop.

The weight of his choices—fight Shell or not—burdened the mayor every day. "What if it is my family...," he sometimes thought, not finishing the notion in words but envisioning two worst-case outcomes.

The first was, What if it is *my family* that lets them drill, and an accident happens, a spill or explosion, and whaling is destroyed, and our culture? *What if this is my fault?*

The flip side was, What if it is my family that *stops* the drilling, and as a result the North Slope goes broke?

Could a disaster like the one on TV happen here? Itta thought now, fretting, heading to his office.

But Itta also thought, *My people need income from oil.*

In Anchorage the weather was milder, clear for flying, and 720 miles southwest of Barrow, 52-year-old Pete Slaiby woke in his loft bedroom in his suburban 4,700-square-foot home, eyed his sleeping Brazilian-born wife, Rejani, and padded into the living room past brightly colored Mexican paintings collected during 27 years working for Shell, and a chess set from Syria, a wooden sitting stool from Cameroon, a glass model of a carnivorous pitcher plant from Brunei. He had his morning coffee in his combination jazz-listening room and well-stocked library on the ground floor, with its huge windows overlooking the gorgeous ice-covered Cook Inlet and Chugach Mountains.

Slaiby had been feeling stymied recently and thinking, *We have to sink those wells.*

The view outside often included Alaska's famed wildlife. Once Slaiby had walked into the room at midday and found it so dark that he was baffled until he realized that a bull moose stood inches away on the other side of the glass, blocking the sun.

An engineer by training, the former sport drinker and wild partier was now a happily settled family man who headed Shell Oil's Alaska Venture. He was the public face of the corporation's Arctic plan, a youthful-looking man with a full head of collegiate-style brown hair, an expression that could be shyly goofy when he was embarrassed, suddenly fierce when he was mad, or quiet and inwardly directed when he was frustrated, which he often was when it came to the North Slope.

He had just under 100 people working for him up on the tenth and thirteenth floors of the Frontier Building—lawyers, political liaisons, drilling experts and engineers—all laboring to sink exploratory wells in the Chukchi and Beaufort Seas, so far unsuccessfully.

Shell is a Dutch company. Slaiby reported to Shell's North American headquarters in Houston, and Houston reported to the Netherlands. Slaiby had an engineer's confidence in technology. Asked about oil spills, he answered, with conviction, "I believe we can handle anything that comes up."

Slaiby's joy that morning came from looking down at his sleeping infant son, Teddy, only a month old. The couple had wished for a child for a long time. The late pregnancy seemed like a miracle. Slaiby had cut out the drinking. He'd cut out too much sun. He wanted to live for a long time to enjoy his son, and he preferred to stay close.

"But it was important to see the mayor."

Shell had decided that the next big US oil find would be in the Arctic. In fact, the company believed that the last huge undiscovered fields *left on earth* lay north of the Arctic Circle beneath the sea in areas once inaccessible because of fierce weather and treacherous ice. But the poles were warming, and between new drilling and shipping technology and high oil prices the feeling was that once-forbidding northern regions would soon open for business.

Shell wasn't alone in this thinking. By 2010 strategic planners across the globe were greedily eyeing the north. The US Geological Survey predicted that over 25 percent of the planet's undiscovered oil and gas lay in the Arctic; with 29 billion barrels of oil and up to 132 trillion cubic feet of natural gas thought to lie off Alaska's North Slope alone.

So after conducting secret seismic work Shell had paid a total of $2.1 billion to buy leases in the Beaufort Sea east of Barrow in

2005 and in the Chukchi to the west in 2008. Much of the oil—the company believed—lay close to Eskimo hunting grounds.

But five years after the first sale the only thing Shell had sunk into the area was $3.5 billion for the leases, studies, equipment and lawyers. Shell had been stopped each year between 2007 and 2009 by legal challenges brought at first by Itta, then a consortium of national environmental organizations teamed up with a few native groups, then by byzantine federal permitting processes.

To many experts, one of the most powerful individuals among these groups was Edward Itta. That is because the Iñupiat hunter spoke to all parties. He had the ear of the White House. He had a bully pulpit to address the Iñupiat people. As an ally, he could help push the plan through.

"The terms by which future oil will flow to the US will be set in large part by North Slope residents," Mead Treadwell, head of the US Arctic Research Commission in spring 2010, had said. "That's how important Edward is."

"What the mayor does affects every American," echoed Kim Elton, director of Alaska Affairs at the Department of the Interior.

So now it was necessary for Slaiby to try to keep the mayor from getting too alarmed about events in the Gulf.

Up until a week ago everything had gone fairly smoothly for Shell in 2010, considering the complexity of Alaskan oil politics. The Obama White House had announced in March that it con- sidered the Arctic leases—awarded under President George W. Bush—viable. It could have opposed them. Itta had opted out of suing. He continued to try to gain concessions through direct talks with Shell, appeals to Alaska's senators, meetings with federal offi- cials and the ever-present threat that at any time he could go back to court. He wanted Shell to provide exact figures on the amount of air pollution that incoming ships would produce. He was demanding that Shell promise to haul away mud and drill cuttings brought up from the sea bottom during operations, instead of dumping them. He was trying to fund ways for scientists to study the conditions allowing marine mammals to flourish off the North Slope.

But he had not sued.

"I haven't filed anything yet. I'm making up my mind," Itta liked to say. He was considered a mediator who preferred to be in the room when decisions were made.

Slaiby was a Republican who believed the nation needed oil and Shell could provide it. On bad days he saw the opposition this way: "They can throw 75 issues on a wall to try to stop us. Only one has to stick for it to work."

Now he walked downstairs to his three-car garage, chose the charcoal-colored Mercedes and headed toward Anchorage Airport's "millionaire terminal" serving private planes.

His thickly wooded street sat off Old Seward Highway in a mixed neighborhood of modest and expensive homes. The news on his radio was getting worse. Although the *Deepwater Horizon* had been a BP rig, Slaiby knew that all oil companies would feel the consequences of the disaster.

"The Gulf wasn't our problem, but in a way it was."

He reviewed facts as he drove. On April 20 an explosion had ripped through BP's 400-foot-long deepwater rig, 50 miles off the coast of Louisiana. Fire had broken out. Despite alleged defenses aboard to stop a blowout, men had been sliced to death by flying shrapnel, hurled across rooms and crushed beneath wreckage. Others had leaped through a haze of smoky gas into the oil-slick Gulf 60 feet below. Two lifeboats pulled away, leaving ten crew behind.

Of 126 crew members on board, 115 were rescued; 11 died.

By the next day, fires continued burning and cleanup efforts were under way by company and US officials who feared that as many as 336,000 gallons of crude oil—roughly 7,500 barrels—might leak into the Gulf.

On April 22, a second explosion occurred 36 hours after the first, sinking the whole rig. By then CNN was reporting that 8,000 barrels of oil a day were pouring into the Gulf, and four days later independent scientists upped estimates to 25,000 barrels a day. The spill was coming from three different breaks. The only good news was that so far, Slaiby knew, good weather patterns held the spreading oil offshore, temporarily sparing the vulnerable beaches and swamps of the Gulf Coast from damage.

Meanwhile the US Minerals Management Service, which was responsible for regulating safety on oil rigs off the United States, reported that it had conducted three routine inspections of the *Deepwater Horizon* in 2010, the latest on April 1, and found no violations. This did little to engender trust in that agency, whose policies had been under assault for months in government reports and the press.

The *Deepwater Horizon* explosion was about to become the worst oil spill disaster in US history. So it was important for Slaiby and his bosses to try to calm Mayor Itta.

"We had to explain the difference between what we wanted to do in Alaska and what had happened in the Gulf."

Slaiby said, "You ignore Edward Itta at your peril."

The big environmental organizations knew this too, and Itta's staff had received a steady stream of phone calls from them in the wake of the Gulf disaster. "They figured surely the mayor would put his foot down and say *never*, but he was still evaluating," said a top aide to Itta, Andy Mack.

The irony of all this, from an oil proponent's point of view, was that the funds that paid for Itta's efforts came from oil. When Itta went to Washington, when he stayed at a hotel in Alaska's capital of Juneau, when he hired lawyers, the monies that paid for it originated with BP, Exxon Mobil, ConocoPhillips and Shell.

Oil tied together Pete Slaiby and Edward Itta. Both needed it. On that day, when the balance between extracting energy resources and protecting environments preoccupied communities across the US, Itta and Slaiby weighed in as two passionate sides of the issue. Slaiby with confidence. Itta with fear. Slaiby boosting technology. Itta his community. Both men wanted to do the right thing. Both took their jobs seriously. Both had different visions of what their jobs entailed and both knew that the outcome would affect every American, and possibly the world.

Itta said, "It is a high being an American. I know the country needs energy. I have to find a balance."

"My mission is to drill," Slaiby said.

The flight from Anchorage to Barrow takes one and a half hours. Alaska is so big that were the state to be superimposed over America's lower 48 states, the east end would touch North Carolina and the west would brush Los Angeles, as shown on business cards carried by some US Air Force officers assigned to Elmendorf base in Anchorage.

Talking strategy, Shell's president, Marvin Odum; Pete Slaiby and Dave Lawrence, executive vice president of exploration— the soft-spoken man who had made the call as to the size of the potential find off the North Slope—passed over Alaska's Seward Peninsula, over the Noatak National Preserve and the massive mountains of the Brooks Range that run along the southern boundary of the North Slope. Suddenly the jet seemed to be flying much lower but that was because the peaks reached up so high. A world of white passed below when the clouds thinned enough to permit visibility. You could fly for hours here without seeing a town or city, streetlight or road. Below lay an ice planet, with oxbow-shaped bends of tundra rivers demarcated by depressions in the snow, the faint oval shape of tundra lakes the barest drop in land. Down there lived many more wild animals than human beings—caribou herds in the tens of thousands; grizzly bears in the south part of the borough, although they also occasionally came north; wolves and wolverines; musk ox and moose. There were the summer nesting grounds of millions of birds that wintered in Nebraska and Mexico, New York and California. There were deep lakes filled with Arctic char. These were creatures that Edward Itta had grown up seeing, hunting and eating.

"We were concerned that the mayor might backtrack. It was imperative that people of the North Slope understand the difference in risk between the Gulf and the Arctic. The Gulf was not our crisis, but we were going to end up wearing it," Slaiby said, worried that once again, Shell's plan would fall through.

By 4 p.m. the mayor anxiously waited for the oil men in his office, in a new and modern borough office building constructed with oil tax revenue. The structure centered on a well-lit atrium with offices along its periphery and others upstairs. There was excellent lighting and computers, fax machines, a Borough Assembly meeting room. Wall decorations included photos of Eskimo dances and blanket-toss celebrations.

With Itta sat a tall, white-haired and deceptively soft-spoken lawyer named Harold Curran, a key aide, an incisive analyst with a razor mind and quietly ferocious temper. The mayor had invited me to sit in as a writer for *Parade* magazine.

Also in the second-floor office were representations of things that mattered to Itta: walrus tusk ivory carvings showing Eskimo scenes, and a large oil painting of one Iñupiat founder of the borough with a bowhead whale swimming below him as he looked into the future, where Itta sat now. There was a print of a lone Eskimo hunter on an ice floe, cut off, adrift, which Itta had hung to remind himself of the "importance of doing things right," he said.

"It wasn't his fault he was adrift, because nature does what it wants to do. But the other side is, he ended up in the middle of the ocean because he didn't pay attention."

Also relevant to the upcoming meeting were two maps posted in the adjacent conference room. They both showed the borough. The first was a topographical depiction of the land Itta loved. The Slope appeared in brown beside the blue Arctic seas and the white of the rest of Alaska. The northern part of the borough was dotted by thousands of elliptical freshwater lakes, cut by dozens of rivers, bordered by Canada's Yukon Territory on the east and by a village called Point Hope—which opposed all offshore drilling—on the western tip. Tiny black airplanes marked places that received regular air service, like Barrow. Red airplanes—there were more of them—showed communities capable of receiving small planes but that had no regular service.

Denoted also were massive federally designated areas including the National Petroleum Reserve and the Arctic National Wildlife Refuge.

This map showed an American Serengeti, one of the last remaining pristine, wild places on earth.

Which was why the second map was so striking. It depicted the same area divided into perfectly formed squares, as mathematical and sharply drawn as boundaries in a Chicago Realtor's guide.

If the first map invited a viewer to imagine natural history and wildlife, the second was the highlighted threat or promise—depending on how you regarded it—of profit and national energy supply. The squares represented "Resource Development Districts." The first map was bordered by drawings of a wolf and a caribou. The second showed the Alaska pipeline snaking south from Prudhoe Bay on the Arctic Ocean toward the Valdez oil terminal in southern Alaska. Instead of native names for locations the map opted for numbers, as in, "U006S0116E."

It was hard to believe the maps showed the same area.

"I'll work with Shell, but I don't trust them. They try my patience," Itta said.

At 4 p.m. the executives finally arrived, shook Itta's hand, and sat down at his oblong conference table. A ceiling fan rotated slowly. A glass jar held M&Ms and peppermints. The executives wore corduroy shirts, boots, jeans, Columbia fleece pullovers. Not a suit or tie in sight.

The room was quiet except for the hum of an air control unit, the mood one of strained cordiality at first. The men asked about each other's families. The mayor brought up the subject of potential air pollution from the incoming Shell drillship and Pete Slaiby assured him that it would use ultra-low-sulfur diesel fuel. The company was overhauling the exhaust system to reduce emissions.

Slaiby promised that the drillship would not even go into the Chukchi Sea if the ice was too dangerous.

Itta listened but said, "I haven't filed anything yet," meaning a lawsuit. "I still haven't made up my mind."

He added, "This thing in the Gulf of Mexico scares the shit out of us."

Now they were down to the purpose of the meeting.

Marvin Odum was a handsome man of about 50, with a friendly, authoritative air. He told Itta, "It's not our well."

"We've been in the Gulf for 30 years without an incident," added Dave Lawrence.

"Here in the Arctic you have shallower water depth, lower

pressure, a simpler system to control," Slaiby explained. "We're drilling an exploration well, not a producing well, and there will be barriers in place to assure that an accident like in the Gulf never happens."

The room seemed filled with pressure. Potential flash points rose and retreated. An occasional joke alleviated tension.

"The first thing I thought about when I heard of the Gulf of Mexico was you guys," Marvin Odum said. "Not doing it right here would destroy our reputation around the world. It's so sensitive that there's no room for mistakes. This will be our most important operation worldwide."

Itta seemed slightly mollified, but when Slaiby asked him to explain Shell's position to his people, he snapped.

"It's been a week and a half and not one of you guys has come out and announced to the people here, 'Here's the difference between what happened in the Gulf and what we want to do.' The onus always falls on me to say, 'I talked to Shell and they're okay.' That's not good enough anymore."

The executives stiffened. Slaiby offered, "We can't have something here that's difficult for you politically."

Odum assured the mayor, "We'll be taking responsibility for explaining this."

After the oil men left, a slightly calmer Itta sighed. "I'm not trying to stop the oil. But if whales disappear, so will our culture. We rise and fall with the bowhead whale."

Itta's crew was not the one to harvest the first bowhead of the season that spring. The honor went to Harry Brower Jr., one of Itta's neighbors in Barrow, a 51-year-old Iñupiat whaling crew captain who was sitting in his skin boat on the ice at 9:15 p.m. on May 1, inches from the lead—the spot where open water began—looking out.

"There wasn't just one whale out there. There were multiple ones, migrating along the open lead. The whales were on the run."

Harry Brower Jr. is a religious man. He regularly prayed for whales, and suddenly a bowhead surfaced a few feet away.

"We're taught that these animals are listening to us. You respect them. You don't brag. You don't say I. You say *we* take an animal, *we* do it by working together."

The first whale was joined by a second, which noticed the men watching. As it tried to escape under the ice its movements pushed its companion closer to Harry, who had never before had the honor of harvesting the first whale of a season. Rising, the powerful man threw the harpoon, which remained in the whale.

To imagine what happened next you have to understand that although most Americans think of a "harpoon" as the entire long spearlike apparatus thrown at a whale, it is only the steel-tipped arrow attached to the wooden shaft. Inside the shaft is a "darting gun" or firearm. If the harpoon embeds itself deep enough in the whale, the animal's outer skin hits a plungerlike trigger. The trigger fires an explosive shell—a hand-packed "superbomb"—into the whale. The wooden shaft then falls into the ocean like the first stage of a rocket ship, to be recovered later since it floats, and the barbed harpoon remains in the whale, attached by rope to a float that marks the animal's location if it dives and heads away.

Harry's superbomb instantly killed the whale.

The massive animal rolled over and began sinking under the ice, dragging the small boat with Harry inside it into the water. His crew tried to hold on to the line that was playing out but they were pulled toward the lip of the ice. The men shouted for Harry to let go of the line but he refused.

Eventually they retrieved the boat onto the ice.

As always happens when a whale is landed, word now went out by radio and cell phone, reaching other camps and people in town. Dozens of men, women, and children rushed out to Harry's camp to help.

"We were anxious to start cutting up the whale, getting it to our community members."

A steady stream of helpers dragged the whale onto the ice and cut it up, a process that ended in the dark by 1 a.m.

"A lot of people were happy. They were going to have fresh food. Catch of the day!"

Each man in Harry's crew and every helper would get a share. Much of what remained would be carved up and saved in frozen cellars dug into permafrost around town. It would be distributed at feasts that occurred over the rest of the year, in churches or the high school during important events, and to needy elders. As tradition dictated, much of the meat was transported by sled to Harry's home in town for an immediate celebratory feast.

Harry and other men hauled in ice from nearby lakes for cooking. "I don't use faucet water. It has a chlorinated taste," he said. The wives set to work boiling meat. Portions were divided into squares that would taste faintly like pot roast, and also into muktuk, a vastly loved delicacy composed of strips of skin with blubber attached.

Between twelve and fifteen tons of meat came from the whale.

"We boil all the parts. The tongue. Portions of the heart. The kidneys. A glorious time, to feed the people!"

Harry also helped cook a fruit drink of plums, peaches and raisins—chilled and served as dessert.

When the food was ready, Harry hoisted his crew's small flag over his house—each captain has his own, made by his wife—and at that signal almost instantly people began streaming toward the house from all over town. Snowmobiles and vans pulled up. Kids arrived by bicycle. Neighbors walked over. The living room floor was covered with cardboard sheeting and the meat lay in pans on the table. Every inch of space seemed packed with happy men, women and children eating. Harry—a rugged, powerful man—hugged each guest.

"There's nothing to compare this happiness to except taking a whole bunch of children to McDonald's," he said. Barrow lacked a McDonald's, but Harry had taken his kids to them in Anchorage.

As people departed they were urged to take away plastic bags filled with muktuk, and like guests at a wedding in the lower 48 gathering up table floral arrangements, happy Barrowites stuffed bags of food into parka pockets.

Among the guests was Mayor Itta.

"We didn't talk about Shell that day. Just whaling."

They were indulging in a tradition that had gone on in Barrow for over a thousand years, one that as community leaders they had dedicated themselves to continuing. One that rapid changes across the Arctic—they both knew worriedly—could end during their lifetimes.

On May 27, 26 days later, as the *Deepwater Horizon* continued leaking, Secretary of the Interior Ken Salazar suspended all offshore drilling in the Gulf of Mexico and Alaska. The stoppage would last until federal officials determined it safe to begin drilling again, he said.

With Salazar's announcement Shell's hopes were dashed for 2010, but for Slaiby and Itta the stage had been set for upcoming battles in 2011.

Slaiby was determined to drill. Was it too much to ask, after paying over $3 billion, that a company might be allowed to *explore* leases it had purchased?

Slaiby was known inside the company as a closer, a fixer. If he was going to keep that reputation, he needed to drill. He thought of himself as a production guy. He *liked* to drill.

Edward Itta had breathing space to seek more protections for his people. But the next year would be his last in office, last chance as mayor to influence the oil giant and Washington while still struggling to guarantee oil revenue for the North Slope. After that someone else would lead the North Slope and Itta would spend the coming years living with the consequences of what he'd achieved or failed to do.

The pressure was growing again. The fact was that *lots* of threatening changes were coming to the Arctic and they did not all relate to oil.

By the time Edward left Harry Brower Jr.'s house that day both men knew that offshore oil was just one crucial issue in their world, that 2011 would be a bellwether year for them, for Shell, for the Arctic and the nation's ability to deal with rapid changes there.

Itta summed it up.

"We were here before oil. We will be here after. We're the ones who will bear the brunt."

CHAPTER 2

The Big Thaw, May

Iñupiats of Alaska's North Slope speak of ice with the respect that a Sudanese hunter has for a lion, the knowledge with which a prize-winning Iowa farmer regards a field of summer corn, the love that a Vermont poet has for New England leaf variations in October.

Ice appears in autumn like the first snowflakes in the Rockies and departs each spring with the arrival of the whales.

The Yupik Eskimos of St. Lawrence Island—in the Bering Sea, southwest of Barrow—have almost one hundred words for ice, not just descriptions of its physical properties but advice for young hunters who learn from an elder.

Qateghrapak is packed soft ice, perilous to walk on. From a distance it appears as pure white. It is slush that has frozen on top, but the lower section remains mush.

Watch it, the word means. You may fall through.

Kiivnin is snow that sits in the ocean so don't mistake it for ice, young hunters. When snow falls heavily, *kiivnin* forms clumps on the water and slightly below. A boat filled with hunters may have trouble pushing through.

Nutemtaq is friendlier, older ice floes that have flattened and are thick, solid. They look as if they've had a layer of snow on them for a while. Excellent platform for hunting!

On *Nunaavalleq,* where walrus have rested for several days, ice is stained brown from their bodily waste.*

Ice words quantify friendliness or danger. Most feared in Barrow is an *ivu,* an almost living incarnation of massive, fast-moving ice. An *ivu* can form suddenly when large masses of ice collide and the momentum creates mini-mountain ranges. A Barrow-based scientist, taxi driver or teenage couple driving north along the shore road on a winter night, heading out of town to view the Arctic's famed Northern Lights, might gaze seaward and see snow-dusted ice stretching with deceptive calm to the horizon.

But the next morning the same view might be blocked by thirty-foot-high ice mountains that move steadily across the beach, even onto the road. *Ivu!*

Ivus can break up whaling camps in seconds and send hunters scurrying for snowmobiles to escape. They can block a hunter from heading home or bury a man too slow to get away. After large waves eroded bluffs in Barrow one winter recently, residents spotted forms of human bodies floating in the sea. They turned out to be a family that had been buried by an *ivu* centuries ago as they slept in their sod home. The eroding bluff had spewed out the corpses, perfectly preserved down to pebbles found in a girl's stomach. She'd eaten them to keep gnawing hunger at bay the winter she died.

Western scientists are continually humbled by Eskimo knowledge of ice. The National Weather Service added the word "ivu" to

---

* Yupik ice terms are described in *Watching Ice and Weather Our Way,* a superb book by St. Lawrence Islanders working with the Smithsonian Institution's Arctic Studies Center.

its more limited vocabulary of terms in 2000 to supplement broadcasts for shippers. Until then there had been no English words to explain the phenomenon. "Fast-moving ice?" "Ice mountains appearing out of nowhere?"

Ivu—implying a whole set of behavior—is easier to say and more expressive.

One East Coast–born biologist based in Alaska who studies marine mammals told me a story to explain Arctic ice behavior and also why elders are revered in Eskimo communities as repositories of ice knowledge.

"I went out with hunters in a small boat. They brought along a little old man who never spoke. I figured he was someone's uncle…as in, you know…old uncle Joe, who you always have to drag everywhere. Every family has an uncle Joe. I didn't think there was a purpose in the guy being there. I figured the young men were just being nice.

"That night we camped on an ice floe. Uncle Joe hadn't said a word all day. The hunters got a stove going and were making dinner. I went behind an ice boulder to take a leak. It was a beautiful night. Suddenly Uncle Joe spoke, so softly that I barely made out the words. He said, '*We have to leave right now.*'

"Well! I've never seen people move so fast. Nobody questioned him. Nobody made a face or looked around. They grabbed that equipment and jumped into the boat and I almost missed leaving, and as we pulled off an enormous ice mass came out of nowhere, crashed into the spot where we'd been camping and destroyed it.

"I thought," the scientist added with awed respect, "*So that's* why they took Uncle Joe along."

Ice is the paramount feature of the 5.5 million square miles above the Arctic Circle. It is the source of stories, legends and history. Front yards in Barrow are living museums for people who value ice. They are strewn with snowmobiles, wooden sleds, snowshoes, drying racks for polar bear or caribou fur, yellow mechanical snowcats, walking sticks for probing sea ice for thickness, fishing nets to be lowered through gaps in ice.

Inside, home computers are turned to the National Weather Service for ice and weather reports. Wives on cell phones ask relatives if a husband or son has returned from a trip on the ice. Families sit around at dinner eating caribou or musk ox stew or smoked salmon and the talk may be of an upcoming journey to Anchorage by SUV, but not on a road or even on land at all in stretches, because the trip involves using a GPS system to drive a new Ford SUV *directly over sea ice* to Prudhoe Bay and then on a man-made ice road south to visit a cousin or go shopping.

Eskimo dances are often mime shows set to drumming and can depict hunting polar bears on ice, whales emerging through ice holes, or lovers journeying over ice, building a snow wall to protect themselves from fierce winds at night.

No written Iñupiat language existed until a century ago. Knowledge was passed on verbally. This is one reason why spoken statements are still regarded more carefully than in the lower 48. People who say something mean it. In a society where a mistake in phrasing can mean death, if you casually tell someone, "I'll give you a call," you better do it or the person will remember you as a liar. Words matter. Poetry emerges spontaneously and often describes the Arctic's fundamental building block—ice.

One day, for instance, I was driving along Barrow's coast road with Richard Glenn, a former whaling co-captain for the Savik family crew. He was also a member of the North Slope Borough Planning Commission, a geologist with a master's degree from the University of Alaska in Fairbanks, a member of an Eskimo dance group and a rock band keyboard player for "The Barrowtones," who play every Saturday night in the roller rink, for free. Come one. Come all.

Richard has so many talents that Pete Slaiby once told him, "I finally figured you out. You're a Renaissance man. You love everything."

He certainly loved ice. As a whaler he studied it offshore. As a caribou hunter he had familiarized himself with its variations on the tundra. A father of three daughters, he knew, looking outside any winter morning, the best way to dress children for ice. Richard the geologist regarded ice tectonics like rock. As a member of the US Arctic Research Commission he still often traveled the world, speaking about ice in venues like New York City's Museum of Natural History.

He was—in a way—an internationally recognized spokesman for ice, and that day, in his truck, he became Richard the poet, whose talent with words was sometimes sought by Mayor Itta at high-level meetings. Even in a verbal society Richard had a special way.

He glanced out at the sea ice.

"One fall I took a twelve-mile hike along this road. I saw open water when I started and ice when I came home. It was the beginning of ice coverage that year. After that day I knew it would change like a living thing, during colder periods growing faster, in very cold temperatures going rigid. During warmer periods it would be loose and weakly cemented, but for the rest of the year it would thicken. Tides and storms come and go. The ocean breaks against itself and forms mountain ranges and now you have plate tectonics on a smaller scale. *In Barrow you see it happen.*

"And then there's a change. Snow drifts on top of it and things warm. You smell it. Do you know how men and women go through 'the change' in their lives? Well, the snow that was on top all winter melts, percolates and drops through. It flushes salt out of that ice that used to be saturated in brine. The ice gets weak and rotten and finally you can't wait for it to break up. It's no longer the strong, vibrant thing you knew.

"That is late spring for us," he said, smiling, "when the whales, seals, ducks and geese come. We wait for open water. A lot happens in life with that time of change."

That's the way it was supposed to happen. That's the way it had happened for thousands of years. But as Richard Glenn and Mayor Itta knew, by 2010 the millennia-old natural structure of the Arctic was changing.

Throughout northern Alaska, Canada, Greenland, Siberia and northern Norway, even the elders were saying that something unprecedented was happening.

The ice—the once dependably permanent feature of their world—was melting away.

While some US scientists still argued in 2010 over *why* the earth was heating up and whether human influence or natural factors were responsible, no one denied that temperatures were higher. The Arctic was warming twice as fast as the rest of the planet, according to the 2004 Arctic Climate Impact Assessment, a study commissioned by the Arctic Council, a sort of mini United Nations for countries having lands within the Arctic Circle: the United States, Russia, Canada, Denmark, Norway, Sweden, Greenland, Iceland and Finland.

The Council exists to foster cooperative decision making, so it monitors conditions. Arctic summer sea ice shrank by nearly 40 percent between 1978 and 2007. Winter temperatures have dropped by several degrees Fahrenheit from a few decades ago. Trees have spread into tundra. In 2008 a wildfire broke out north of the Brooks Range in the North Slope, in a place where the local dialect had no word for forest fire.

"We've never seen ice melt like this in history," ice forecaster Luc Desjardins of the Canadian Ice Service said.

By 2008, when Pete Slaiby came to Alaska, even officials in the administration of President George W. Bush, which adamantly opposed embracing any idea that human influence has a substantial role in earth's climate, fretted about a hotter Arctic.

That summer I stood aboard the US Coast Guard cutter *Healy* north of Barrow as a helicopter landed on deck carrying Bush's secretary of homeland security, Michael Chertoff, and Coast Guard commandant Thad Allen. Wearing crisp blue uniforms, they awarded the crew a commendation for work furthering US economic and strategic interests, "in the face of climate change as the ice recedes further each spring," it read.

The next day, at 6 a.m. on the bridge, Chertoff told me, "I don't know why the Arctic is warming, but it's a fact and we better get ready for it."

Allen added, "All I know is there is water where there was once ice. And where there is water, the Coast Guard is responsible for it."

They were not talking about only oil and not only about environmental issues, although like millions of people they knew that Arctic ice helps cool the planet, that its disappearance could speed warming, that animals like polar bears and walrus could suffer as ice cover diminishes, that melting freshwater glaciers in Greenland might shrink the percentage of salt in seawater and raise ocean levels.

But Chertoff and Allen referred to *national security* issues, and several loomed as ice melted. The Northwest Passage—the long-dreamed-of trade route between Europe, Asia, and the US, around the top of Canada and Alaska—could open to ships in summers as soon as 2020, some computer models predicted.

If that happened, up to 25 percent of earth's shipping might be passing Barrow within ten years, and if the specter of one drill rig could bother whalers, the idea of hundreds of unregulated ships out there was a nightmare.

"The Bering Strait could be the next Panama Canal," predicted Mead Treadwell.

Why? Money! "A single Chinese container ship sailing between Shanghai and New York could save up to $2 million on fuel and fees each way, using the northern route instead of the Panama Canal," Scott Borgerson, Oceans Fellow at the Council on Foreign Relations in New York, had told me.

The Coast Guard, however, had no permanent base in the Arctic. No way to monitor ship traffic or know whether or not a vessel was friendly, or about to rupture and spill oil, or whether it carried proper lifeboats for passengers.

"We're going to have problems," said Admiral Allen, who kept bringing Washington VIPs to Alaska to push for more money for Coast Guard needs.

Chertoff and Allen also referred to a treaty the US was considering ratifying that would allow any country on earth with ocean frontage to claim new undersea territory and mining rights offshore. North of Barrow this could mean a US gain the size of California.

The secretary and commandant knew that *because* of this treaty—the United Nations Convention on the Law of the Sea—a race had broken out among Arctic nations for control of undersea territory. Some claims overlapped. Russia wanted title to an area the size of France and Spain combined. Some countries sought to block the claims of others. Some were sending troops to the Arctic as their leaders spoke of possible conflict.

"The Russian claim," said Ariel Cohen of Washington's conservative think tank the Heritage Foundation that year, "is a time bomb."

There are remote places on earth that remain relatively unknown in one century and become pivotal the next; places never mentioned in headlines that one day suddenly fill them. The Arabian Peninsula—considered a wasteland of desert—transforms into Saudi Arabia, the oil-rich *kingdom*. The malarial jungles of the

Isthmus of Panama are gouged out to form the strategic and critical Panama Canal.

The North Slope of Alaska could be such a place in the 21st century.

The bottom line in 2010 was that the Arctic was still harsh and dangerous but not inaccessible. It was breaking into spheres of national influence. Those who controlled it would prosper. Those who lost opportunities or failed to prepare would gnash their teeth.

Rudyard Kipling's "Great Game," his term for the jockeying among global powers for influence in remote regions, had reached the far north.

Borgerson predicted, "In twenty years the Arctic coast of Alaska may look like the Coast of Louisiana today, lit by the lights of ships and oil rigs. One port there may become a trade hub as important as Singapore. Singapore, once a mangrove swamp, is now the biggest seaport in the world."

I'd sent out a flood of e-mails asking scientists, naval authorities and diplomats if there might be one place that epitomized the big issues facing the Arctic, one place where Eskimos had lived for thousands of years where I could find oil exploration, climate science, potential for conflict, shipping opportunities and undersea land claims.

I was surprised when the same answer kept coming back. As John Walsh, director for the Center for Global Change at the University of Alaska, said, "Barrow is Exhibit A."

I was also advised by Willie Hensley, Iñupiat Eskimo author and former Alaskan legislator, to visit Eskimo villages farther south in Alaska, along the Bering Strait and just below the Arctic Circle or just above it. Changes predicted for more northern latitudes had already occurred there, he said. What I saw in these villages would not be computer predictions of future change but actual sights.

Accordingly, in the summer of 2009 I boarded an Army Reserve Black Hawk helicopter in Nome and with a contingent of US Army and Air Force doctors headed out for ten days of visits to remote villages. The Coast Guard's "Operation Arctic Crossroads" was designed to familiarize staff with operating conditions in the north. But the Coast Guard included a humanitarian element. After asking village leaders what their communities needed, they'd decided to hold free medical clinics and provide veterinarians to vaccinate dogs. Rabies is widespread in the Arctic.

Each day we flew hundreds of miles. Nome lies below the Arctic Circle, and the landscape nearby was spectacular and mountainous—forested in places with magnificent cliffs lining shores, snowcapped peaks that often seemed only feet away as we twisted through deep gorges, and vast tracts of boggy lowland where sun reflected off rivers and swamps as thick as the Amazon's. It was a world of water. We saw herds of musk ox and grizzly bears gazing back or fleeing into thick forest, dall sheep on mountain crests, eagles soaring, moose grazing in lakes.

We saw no people from the air but occasionally spotted trails made by four-wheel-drive runabouts.

Each village had a story and it was always about ice.

In seaside Kivalina—a tiny settlement clinging to a barrier island on the shore of the Chukchi Sea—we found Army Corps of Engineers bulldozers piling immense boulders into barriers lining the shore. Sandbags surrounded the power station like a handbuilt levee protecting residents of a Mississippi River town.

Mayor Colleen Swan explained that shore fast ice has disappeared in recent years. It no longer protected the land when fierce autumn storms pounded the settlement. "Waves are bigger and hit harder. We're going to have to move the whole village or we will be overwhelmed."

In lovely Koyuk, which sat high on a forested bluff overlooking a meandering river of the same name, tribal head Merlin Henry took a day off from reindeer herding and rode me around on the back of his runabout to visit elders. It was the first day of moose-hunting season. As we went from house to house, so did hunters carrying bloody sections of meat. They'd deposit it in kitchens as a gift and leave.

Elder Robert James, who had once been a National Weather Service volunteer observer, told me, "Sea ice is melting sooner in autumn. Flies are thicker in summers because of heat and make the caribou sick."

People spoke of spruce bark beetles—never before seen this far north—falling from the sky like rain. Their infestations explained why the helicopters had flown over miles of dead forests. Inland, Merlin said, when hunters went after moose, their boats increasingly ran aground in flats as the banks fell in, and frozen tundra was melting and lakes that had been there for centuries were drying up.

"It's harder to find food," people said.

A few days later, on St. Lawrence Island, in the Bering Sea—a 90-mile-long collection of mountains and tundra—I met 76-year-old Chester Noongwook, last surviving dogsled mail carrier in the United States, now retired. Chester had co-authored the book *Watching Ice and Weather Our Way*, which records observations of the natural world.

Conditions have changed so much, Chester acknowledged as we sat in the Savoonga village tribal headquarters building, that some elders now doubt their knowledge of ice behavior and have grown reluctant to pass information to younger hunters in case they make a mistake and cause someone to fall through ice and get hurt.

"The world is spinning faster now," Chester said, meaning that weather and ice had become less predictable.

It was sunny, and afterward Chester's cheerful son Milton, 49, a former secretary of the local tribal council, showed me around the village. On the way he pulled out a Sibley field guide to birds of North America. So many new kinds of birds are showing up, he said, that villagers need the book to identify them.

Milton then pointed out a series of large wooden boxes set into a seaside bluff. He pulled open the top of one and I saw it was a door leading down by ladder to a cellar carved into permafrost. Milton's flashlight beam illuminated hunks of meat amid a sheen of frost. For centuries, he said, Eskimos had stored food year round in these natural freezers.

But I saw trickles of water down there too.

"It's melting," Milton lamented. "It never used to do that. If it gets too warm, the meat will spoil." Nome's mayor had also told me about melting permafrost, saying it was buckling beneath roads, bridges and buildings.

One day a contingent of Washington VIPs from the newly constituted Obama administration joined us on a trip. They included David Hayes, powerful undersecretary at the Department of the Interior, which awards oil leases and drilling permits on the US outer continental shelf; Dr. Jane Lubchenco, who ran the National Oceanic and Atmospheric Administration (NOAA), which has authority over several permitting activities involved in oil exploration too; Heather Zichal, deputy assistant to Obama for Energy and Climate Change; and Nancy Sutley, chair of the president's new Ocean Policy Task Force.

All of these people would be involved in—or privy to— appraisals of Shell's drill plans over the next couple of years.

The flight itself underlined gaps in Coast Guard Arctic preparedness. Although the three copters flew within several hundred yards of each other, pilots lacked equipment to talk to one another and had to patch through commercial aircraft. We flew by sight, not equipment. The day was wet and cold, the gray clouds thick and low.

From the air, the village of Shishmaref seemed perched on a beach and in danger of going under. The Army Corps of Engineers was building another rock levee there. The VIPs—dressed lightly for Washington—briefly met leaders who told a similar story to that of isolated Kivalina's. Without protective shore ice, the village sat at the mercy of storms.

"We have an intellectual sense of what climate change is, but we're here to put a human face on it," said Jay Reich, deputy chief of staff, Department of Commerce.

"The Arctic is national security. It's wealth. It's geopolitical importance. The president views climate and energy policy as interconnected with the economy," said Heather Zichal. She added, "Until now, the Arctic has very much been out of sight, out of mind."

The villagers were worried. Words were not action.

The VIPs stayed for an hour, then left.

The Iñupiat name for Barrow is Ukpeaġvik, or "the place for hunting snowy owls." From the air, the US Arctic capital appeared as a triangular-shaped settlement hugging the edge of the continent at the junction of the Chukchi and Beaufort Seas. The ocean looked black as anthracite in July and was specked with whitish ice. Snow can fall any day of the year in Barrow.

I'd flown in on one of two Alaska Airlines 737s serving the city each day and that carry cargo in front, passengers in back. You enter and exit planes through a rear door and use mobile stairs to reach the tarmac. The walk to the cramped metal terminal can be freezing in winter, when temperatures average 16 below zero and can drop to 70 below. In summer they average 46 degrees.

Barrow's famed polar day, when the sun never sets, lasts about 80 days from late May to late July. Round-the-clock darkness runs from late November until late January.

If out on the ice what was threatened by oil development was evident—surfacing whales, seals resting by holes, pristine waters and polar bears on floes—then on the streets it was what oil revenue *paid for* that stood out; simple sights taken for granted in any American town.

Driving around in a pickup truck, I saw a new hospital going up and plenty of cars, runabouts and taxicabs on miles of unpaved roads that crisscrossed the city, and every vehicle I saw had windshields cracked by flying gravel. Office buildings—for the borough, police department, Wells Fargo Bank and Arctic Slope Regional Corporation—stood near one intersection. I passed a convenience store and a variety of restaurants: Arctic Pizza; Pepe's North of the Border, serving Mexican food; Northern Lights, Brower's Café in an old whaling station, where you could dine on burgers, sandwiches, salads, Korean food; Osaka Japanese restaurant, where Mayor Itta liked to eat pancake-and-egg breakfasts sometimes.

Ilisagvik Community College, the only tribal college in Alaska, had its dorms on a refurbished old naval base a few miles away, up a coast road lined by city garages, an auto parts store, even a self-service filling station supplying natural gas fuel to drivers who had converted their cars to run on it.

The normalcy of these sights masked the huge expense of creating them in a place where heavy materials have to be barged in by sea.

The high school that served 200 local students had cost $80 million, for instance. It provided school lunches and had a swimming pool available for use by the community. The drywall for homes cost $43 a sheet in Barrow, compared with $10.82 in Seattle that year.

In spots I noticed what appeared to be small, six-foot-tall wooden shacks by the side of roads. They turned out to be aboveground entry points for Barrow's "Utilidor," a 3.2-mile wooden underground tunnel carved into the permafrost, that brings potable water from a freshwater lagoon, sewage service, cable TV, fiber optics and telephone lines to homes. Temperatures are kept at 48 degrees in the tunnel, and the project cost over $300 million to build.

Mead Treadwell had told me that HUD, the US housing authority, had refused to install plumbing in rural Alaska public housing while building it in the early 1990s. The cost was prohibitive, Treadwell—then Alaska's deputy commissioner of environment—was told. Residents could have walls and heat, just not toilets. The North Slope Borough had to chip in for the plumbing.

Just a stroll through the warehouse-sized main store in town, the AC Value Center—where you could buy fresh fruit, DVD movies, new snowmobiles, steaks, undershirts or parkas—showed the difference in prices between the North Slope and the lower 48. A gallon of milk cost $9.99, not $2.49 as in Seattle. A 28-ounce jar of Skippy peanut butter—$3.99 in Seattle—in Barrow cost $8.95. Food was expensive because, like the building materials, it had to be barged in by sea.

And Barrow had it easier than outlying villages did. There gasoline could cost up to $8 a gallon and store-bought food cost much more.

What Mayor Itta sought to preserve across the borough was more than the ability to purchase consumer goods. It was a whole lifestyle. There was a well-stocked, well-lit library that sat beside the heritage center, a museum dedicated to explaining Iñupiat life, where local whalers built their skin boats in a large workshop out back, local artisans crafted statuettes of walrus ivory and carvings of baleen and the mayor worked with the Rosetta Stone company to create educational tapes teaching Iñupiaq to youth. (Iñupiaq is the Iñupiat language.) There was a fairly new, comfortable home for the elderly, with wheelchair ramps and clean, well-heated rooms. Hallways there celebrated local history and were lined with black-and-white photos of elders when they were younger, often out on the ice. Large trash bins on corners displayed slogans pro-moting "Healthy Communities," a program that Itta had created to convince young people to stay away from alcohol and drugs.

The city's diversity was growing. People you'd bump into in the Value Center included lawyers that the borough had brought in to safeguard community interests, or scientists from around the world working on climate and ocean issues. There was no movie theater but there was a roller skating rink; no bowling alley but there was a blue Astroturf high school football field. For a commu-nity of 4,500, Barrow had a rich mix of cultures and things to do.

Back at the airport I ducked into a shiny twin-engine Beech-craft King Air operated by the Borough Search and Rescue Squad, which employs ten pilots who also fly a Learjet 31 and two Bell 412 helicopters. They carry out 250 medevacs a year throughout the 89,000-square-mile North Slope, Director Hugh Patkotak said. They retrieve injured hunters or sick villagers and, if the situation is dire, deliver them to specialists in Anchorage at no charge.

The squad's proudest moment occurred several years ago, when it saved 164 whalers after an enormous field of ice that they'd camped on broke off and "they were drifting out to sea," Patkotak said.

Working all night in poor visibility, the pilots landed again and again on the ice, evacuating five men at a time, then snowmobiles and sleds and tents.

Hugh was a tall man with a friendly disposition. In the spotless hangar he gazed at the shiny copters.

"This all goes away if oil goes away."

By spring 2010 many offshore oil advocates and opponents could agree on one thing at least: the United States was notori-ously unprepared for changes occurring in the region. The nation had not signed the Law of the Sea Treaty. It had not filed a claim for territory. The US had one functioning icebreaker to address emergencies, while the Russians had twenty. The US lacked proper communications equipment, lacked a deepwater port, lacked even basic science that could inform decision makers as to natural processes in the region before they made plans. There was no cohesive national policy for addressing Arctic energy extraction.

In contrast, other Arctic countries—Russia, Canada, Norway, Denmark—were much further along when it came to purchasing new icebreakers and awarding undersea oil and gas leases, and they had begun the process of expanding their national territories through the Law of the Sea Treaty.

"The Arctic is crucial and the Arctic is now," said Adm. Gene Brooks of the Coast Guard. "But if this were a ball game, the US wouldn't be on the field, in the stands, or even in the stadium."

"We are at a critical time in the Arctic," Alaska's Republican senator, Lisa Murkowski, told a round table at the Council on Foreign Relations in New York. "We can go down one of two paths, one of conflict, the other of diplomacy."

"Every day that goes by we lose our sovereignty," said Alaska's Democratic senator, Mark Begich.

In short, what would happen in Barrow wouldn't stay in Barrow. It would ripple across the nation and world.

At a time when oil prices were rising, the economy slowing and security challenges multiplying, the fate of Shell's drilling plan over the next year would be a bellwether for the larger question of whether the US was dealing with or oblivious to what an opening Arctic meant for the nation's future.

Pete Slaiby said, frustrated, "We're an oil and gas company. We believe oil and gas is in everyone's best interest and will be a hugely important part of the energy mix during the next century."

Mayor Itta said, caught in the middle, "We live at ground zero for climate change."

CHAPTER 3

Edward's Dilemma, June

Alaska was a US territory—not a state yet—when Edward Itta was born in 1945. Barrow was populated by 360 people. Homes lacked running water. There were no roads. Travel was by dogsled. Plastic pails called honey buckets served as repositories for human waste. In poorer homes they sat in the open. In better ones like Itta's they had their own room.

"We burned blubber for heat. You don't have to tell me how important oil is."

The Bureau of Indian Affairs ran the only school in town, conducting classes as far as grade six.

"We did not know we were poor. We had food and clothes. I don't remember suffering from cold. There was a sense of being well taken care of," he recalled.

Edward's father, Noah, fed his family of ten children and his eighteen dogs by hunting and carpentry if there was paying work. His mother kept a house that old friends recall as filled with boisterous relatives and neighbors.

"If I tried to leave at dinnertime, Edward's mother would call to me, 'Where are you going? It's time to eat,'" childhood friend Lloyd Nageak recalled.

"We were never like a traditional Western family, as in, *me. My* wife. *My* children," said Itta. "*Family* meant your whole extended family; grandma, cousins, uncles. If you went to see a cousin, you stayed there. If you did wrong, you were scolded by any adult. And we shared. My father had a skin boat and dog team. My first cousin and his nephews were welcome to use the equipment, guns and bullets, and expected to share whatever they got with us."

The best memories involve the outdoors. In summers the family would climb into a boat and go upriver, camping. Itta's first duck hunt. The first time he tried to skin a caribou. Running on a beach with his sister Dorcas one glorious summer day and learning how to catch fish in a net. Falling asleep next to his favorite uncle, George, and being woken the next morning with a kick on the feet. *Hey! Get up! There are ducks flying! We gotta go!*

At the same time, the pressure was starting, the need to do well. Noah Itta had "a hard time accepting failure in his kids," said Mark Wartes, another old friend.

Edward at six years old could not foresee that years later he would be described by admirers as having one foot firmly in the Iñupiat world and the other in the modern US. Sen. Mark Begich would call him a "practical compromiser who doesn't forget his roots as an Alaskan native and who has done a lot to bring parties together." But the dual schooling—for better and worse—began early. Barrow had been a contact point between Eskimo and Western cultures since 1852, when the British exploring ship HMS *Plover* arrived and wintered over for two years.

Initial contact was peaceful. Then Western whalers seeking bowhead oil for fuel arrived around 1848 and introduced modern equipment like the exploding whale bomb, a cash economy and ideas like camping out on the ice during hunts. Even then the North Slope meant energy to outsiders.

When more easily obtainable oil was found in the ground in Pennsylvania in 1859, the Yankee whalers came north to harvest baleen, a substance that lines the mouths of bowheads and filters out crustaceans for feeding. It was used in the manufacture of women's corsets.

Two thousand pounds of baleen—from a single whale—could fetch $11,000 on the market by 1888, an enormous sum.

Native and Yankee whalers taught one another hunting techniques. Some white whalers married into villages, where anyone who was willing to pitch in on a hunt was welcome. One positive aspect of contact for local Iñupiat whaling captains was that their prestige was enhanced as they used their share of baleen money to outfit crews.

Traditional religious leaders, shamans, on the other hand, saw their influence plunge in the face of Western medicine and missionaries. Life was changing fast.

If cash arrived with outsiders, so did alcohol and disease. Their decimating effects were chronicled by Charles Brower, a New Jersey–born whaler who settled in Barrow, married an Iñupiat and established a whaling station that exists today as a restaurant.

In his superb book, *Fifty Years Below Zero*, Brower described one gruesome flu outbreak in the 1890s. It struck inland Iñupiats who'd come to town to sell furs to whalers who "stayed just long enough to give them some kind of germ."

After the ships left, the inland Iñupiats, already growing ill, loaded their boats for the trip home. Brower hoped they'd get better on the way, but soon afterward a hunter rushed into town to report dead bodies scattered along a river. Brower went to investigate and wrote:

Following the homeward course of the stricken people we came to the first evidence of disaster not ten miles away—a woman with a young baby...

The farther we went, the greater grew the tragedy. It was clear that from the day they set out for home they had been dying all along the coast and up the rivers. From the postures of the bodies, we could almost visualize it happening—the stronger members dragging the weaker on to the banks to die, then paddling a little further until it came their turn to be abandoned...

All that fall and the next summer we kept getting reports from hunters of bodies discovered along far river banks, sometimes alone, sometimes with a few belongings scattered around.

Another influence that Brower chronicled was Christian missionaries who urged Eskimos to move to towns and live near churches, give up nomadic ways and stay put.

But the Iñupiats lacked materials to construct solid homes. They ended up packed into rickety shacks that were freezing during brutal winters, when temperatures dropped as low as 70 below zero. Cheap wood-frame houses could not retain as much warmth as igloos or traditional sod houses. With no running water in homes, diseases flourished.

Brower wrote, "The only thing to benefit from the 'new order' was the consciences of the well meaning missionaries themselves."

Many Eskimo village populations had plunged as much as two-thirds from 1840 levels by the time Edward was growing up.

Like a small planet pulled into the gravitational orbit of a larger one, the traditional Iñupiat world had, from first contact, been shaken from its axis. The contrasting demands of two cultures would tug Edward in opposite directions all his life. Early on they made themselves known in as small a place as a Barrow sledding hill.

Lloyd Nageak recalled 60 years later, "The kid running the hill said, *only pureblood Iñupiats can use this hill.*" Since Edward's grandfather had been a Portuguese whaler, making him one-quarter white, "he trudged off. I went with him."

The bigger shocks started in elementary school. There, students were forced to speak English, punished for speaking Iñupiaq. But at home English was discouraged. If you spoke English, "Noah would just give us a look," said Mark Wartes, "and we'd stop."

He added, "Our third grade book had a little red-haired boy in it named Patrick. He lived in New York City in a high-rise and grew a pumpkin on his windowsill. There was not a thing in that book that we could put together with our lives. We didn't know a Patrick or kids who had red hair. A high-rise? What was a pumpkin? How do you grow something in a window? None of this computed."

Unfortunately, the BIA teachers often regarded unfamiliarity with subjects as stupidity, Mark Wartes said.

"We had this teacher from Alabama. We couldn't understand her accent and we were trying to get a grip on what was coming at us. Sometimes you needed a few seconds to translate in your head from English to Iñupiaq. But her attitude was, *I have a whole lot of dumb little kids here.*"

Also, unlike Iñupiat adults, who taught children by kneeling down at eye level, Western teachers stood and looked down at you. To small children it felt as if they were being talked down to, many Barrow elders recall.

Like many in his generation, Edward still remembers his BIA education with gratefulness for the learning but hurt at the experience.

The process that began there—the sometimes subtle, often well-intentioned, occasionally antagonistic and never ending assault on Iñupiat traditional values—would continue throughout Edward's life as government agencies, corporations and environmental groups worked to change Iñupiat mores into ones common in the lower 48 states.

Edward was ten years old the first time he was asked to actually help to do this. One day a stranger showed up in school asking if teachers could spare four boys for the shooting of a film. The four were Edward and his friends Mark Wartes, Lloyd Nageak and Solomon Nummik. The resulting black-and-white production, distributed in churches around the United States, helped raise funds for missionaries. It was designed to show the way Eskimo children grew up in Barrow.

In *Adventures with North American Neighbors*, grainy footage showed the boys cutting ice with hatchets for drinking water, running dogsleds and sliding down hills for fun in their parkas as they owned no recreational sleds.

The boys seemed happy—life exotic—but Solomon would commit suicide as a teenager. Mark—son of a missionary—would be so jolted by moving away from Barrow to Seattle, feeling so alone in a city, that he would disappear into the tundra to spend fifteen years as a hunter.

Lloyd and Edward would learn to drink.

Their stories—depicted as typical of the North Slope—would turn out to be that in more ways than one and would inform much of Itta's attitude as mayor, later, as he dealt with state and federal governments and with global oil companies.

But in elementary school these shocks had not occurred yet. Edward the self-described rebel was "always getting into trouble, questioning things." Of the adults in his life, Uncle George Ungarook, his mother's brother, would have a major lifelong influence.

"George was a bachelor," he said. "Very well read. I think in every family you'll have an uncle who notices one nephew or niece and *man, that kid!* My uncle took me out, took me hunting.

He told me if I goofed up. When I started becoming aware and conscious he always made a pointed effort to take me along and teach me. How to hunt ducks. How to get fish.

"And every so often he would start talking about the ways of the world. That white people had a different view. Our way of life was not going to be what it was anymore. He said I would not understand at first but that we were at a turning point."

They were just words, but they took on more meaning when Itta began seventh grade and was flown to Wrangell, Alaska, to attend a combination Bureau of Indian Affairs boarding school and orphanage for Native American kids. It was the only way to continue school beyond sixth grade.

"I didn't want to leave my family, but my folks made sacrifices. Education was important to them."

At Wrangell, suddenly there were no supervising adults around except matrons—strangers, who left at night.

"I cried a lot. I felt homesick. You knew something was wrong, but we had no word for culture shock. I remember the first time I saw trees. A car. A strange, alien world."

Edward's younger sister Dorcas, who also attended the school, still has nightmares about it. "It was an ex-military camp. We slept in barracks—ten, twelve girls in a room. We were assigned a little child to take care of, to wake up every day. They'd pee in bed and you'd clean them, wash them, take them to the cafeteria. The little Yupik kids couldn't even speak English. They just cried and cried for their parents. The lights would go out at night and one girl in the dormitory would start crying and then we all would. I remember kids who had never been in a shower before. It would come on and they'd start screaming in shock. Later when I grew up I'd still be haunted by their cries."

What would haunt Mayor Itta in 2010 was the specter of losing funding for schools across the North Slope if oil money dried up.

After high school Edward spent 22 months in Cleveland, learning electronics, again from the BIA.

"We got lessons on how to read transit maps. To budget money. You got an allowance, but once it was gone you were out of luck and out of food."

One day, the bus he used to get to school could not get across Cleveland because the Cuyahoga River was burning. This was astounding to an Eskimo. Chemical discharge from factories had so polluted a river that it ignited. Years later, Itta would recall this while watching the *Deepwater Horizon* burning on TV, and trying to gauge the danger of a potential oil spill off the North Slope.

"If we lose the ocean, we have lost the Iñupiat Eskimo," he'd tell reporters, images of those flumes in his head.

By the time the Cleveland years closed out, the boy had one foot in both worlds, all right, but not as comfortably as non-Eskimos thought. The skills he had been taught to value in Barrow—hunting, sharing, knowing how to survive outdoors—had little use in Cleveland. The Vietnam War was "consuming America," and as Iñupiat men have a proud tradition of serving in the military, Itta enlisted in the Navy and worked as an electronics technician on a destroyer in Southeast Asia.

"My job was to help run the communication equipment."

On the ship, crew members often asked if he was Mexican.

"I'd answer and they'd say, you're an *Eskimo? Woooo!*"

He never felt discriminated against, but as the only Iñupiat he felt pulled in different directions. "I don't think anybody knew it. Sometimes I wished, Why can't I be like *those people*," he said, meaning like the folks back home. "Sometimes, why not *these*," meaning the rest of the crew.

The boy who had grown up loving the land, learning to skin animals and catch little silver fish was living in a steel "tin can" that "had six-inch guns. We pounded targets day in and day out."

After six months of deployment the ship went back to base in San Diego. "On TV I saw how unpopular that war was. One day out of the blue, I thought, *I am an Iñupiat Eskimo. What am I doing, killing people? Iñupiats do not go around killing people.*

"That was when I became aware of how the government can run your life if you just follow what they say."

He didn't do anything about it at first but it ate at him throughout the next deployment. "When we came home and nobody greeted us, there was a feeling of being shunned and reviled for being in the military. I decided this war was not right. I tried to become a conscientious objector but didn't qualify. You can't just say 'I object.'"

Itta went AWOL, headed for Canada, where he had relatives, but got arrested while still in California. He was handcuffed and thrown in a brig.

"I was guarded by Marines."

Allowed out of solitary confinement, Itta learned through the grapevine that there *was* a way to get thrown out of the Navy, and it was to take illegal drugs.

"I intentionally set out to get busted."

Itta got caught smoking marijuana. He was kicked out, free finally but also angry, confused and ashamed.

"I was 25, 26, and could do whatever I wanted. I was invincible, but the undesirable discharge hung over me. I knew my family was disappointed."

Thus began a spiral into a dark period. It was a freewheeling moral time in San Francisco, "the summer of free love." Itta was welcomed into the hippie movement. He married a white woman, an artist. Going back and forth to Barrow in the 1970s, trying to figure out where he fit in, "I became part of the Me syndrome. Me against my wife. Me against the world. Me against my society and against the government. I was drinking heavily and into drugs."

As a boy, adults had always chided Edward, if he was mischievous, by saying, "Be a real person. Be an Iñupiat." But what *was* an Iñupiat man in the 1970s? A sailor? A hunter? A killer of evil communists? He was none of those things. He'd lost out on learning hunting skills while attending the very schools his parents had sent him to. Uncle George had ridiculed him over this even though George had urged him to get a Western education. The world made less sense.

And then Itta got a telephone call when Uncle George, his favorite relative, shot and killed himself. George had taken to alcohol.

"I still think about him every day. My idol. My hero."

Many scars, buried deep, would not heal even by 2010. Itta came from a generation where men don't confide problems in wives or children, he said. Asked that year if he was aware of any deep anxiety in his father, Itta's son Price said no, although the men are close. Richard Glenn, in describing the mayor, said, "Edward is in a cadre of special people who straddle two worlds almost without effort. They grew up without modern contraptions. They telescoped into one generation the kind of change that everywhere else in the world takes multiple generations. It's as if our culture got pulled through a keyhole. Famine. Death. Flu. Alcohol. Edward learned to be fluent in two sides. Leaders in his age group are confident walking in two worlds."

Well, he was walking in two worlds, all right, but confidence was still another matter. And by spring 2010 oil was the key to the best and the worst of both worlds.

Like most of his constituents, Edward had no problem supporting onshore extraction, even in federally protected lands like the Arctic National Wildlife Refuge (ANWR). To North Slopers the whole vast Wyoming-size borough contained just as much magnificent wildlife, beauty and uniqueness as ANWR. Outsiders flew in for a day or a couple of weeks to ooh and aah about the protected reserve, but there was nothing particularly different to Iñupiats about it except that it lay adjacent to Prudhoe Bay and so probably contained land-based oil. Extract *that* and you wouldn't have to drill offshore and threaten marine life, they said.

ANWR, to many North Slopers, was a huge area that the federal government had arbitrarily drawn a line around and decreed that you could not take oil from. Edward believed that ANWR was sacred in Washington, that the federal government and national environmental groups would never allow oil extraction there.

Which meant that offshore oil had become the two-edged sword, a devil's bargain: tax income now for a potential disaster later if you didn't build safety into the process.

"It kept me up at night sometimes."

When he'd run for mayor originally, Edward had not thought that oil would be the big issue during his tenure. He was opposed to all offshore extraction then, unlike his opponent, who thought some would be necessary. His big issue was proposing "healthy communities," a fight against problems caused by drugs and alcohol. Borough mayoral candidates do not run as Republicans or Democrats. Elections do not take place on the same day as state or federal elections and are local affairs. It did not occur to Itta in the beginning that once elected he would spend up to 70 percent of his time dealing with oil. Or that he would find himself deciding, as his opponent had argued, that offshore extraction was inevitable. He did not think that one day he might become "the mayor presiding over declining revenues that start blinking off the power plants and shutting the sewage treatment plants in our communities," as Richard Glenn said.

Then again, any notion that a drugged-up Edward Itta would ever run for office would have seemed absurd in the 1970s as he went back and forth between Barrow, where there was no work, and San Francisco, where his wife lived and where there were plentiful alcohol and drugs.

"All I thought was, Where can I get my next fix?"

A slow recovery—connected to Big Oil—began during this period when he heard about a job as a roustabout at the beginning of the Alaska oil rush. The ARCO oil company had brought in the first land-based North Slope well at Prudhoe Bay, 200 miles east of Barrow. The field there was huge. Thousands of people were flocking to Alaska to build a pipeline and explore for oil.

"I'd never heard of Prudhoe Bay. I knew I was working on an oil rig but had not a clue of the implications of oil and what a difference it would make."

For $5 an hour Itta labored on the rig. Later he ran a loader for the borough and worked as director of public works and capital improvements for the mayor he would one day run against and replace.

At the same time, his first marriage failed and his second, to Elsie Hopson, a teacher, provided a solid base for a healthier lifestyle. Elsie gave him a choice. Stop the drinking or she'd leave. Edward credits his escape from the dark times to Elsie.

Still, when depressed or under high pressure Edward in 2010 might leave Barrow sometimes, hole up in a hotel room in Anchorage or Juneau and give in to drink.

He wasn't proud of it. But he admitted it. It was no secret in the community. It had been a campaign issue, as drink was a big problem across the North Slope.

A man at 65 years is shaped by the sum total of his life experiences, and in 2010, millions of TV viewers around the world, watching *Deepwater Horizon* hearings would see an Eskimo named Edward Itta testify via satellite from Barrow. Looking into their screens New Yorkers and Parisians and Chinese viewers in Beijing saw a well-dressed man at a speakerphone addressing the powerful presidential commission with confidence and conviction.

But had they looked into his heart they would have seen a man who was still battling personal demons, haunted by ghosts of loved ones and memories of official opposition to every step in the creation of native power, the fight to establish the borough as an entity, the fight to tax Big Oil, the fight against the 2007 Shell plan—a man who privately feared that he was making the wrong choices even as he struggled to make the right ones and protect his people from the next round of changes affecting them, as personified by Shell Oil and Pete Slaiby.

By June of 2010 Itta had a reprieve from pressure. Shell had pulled the plug on their plans for that year. It seemed ridiculous to spend millions leasing ships and equipment in the face of the Department of the Interior's drilling suspension. The oil giant's goal was now summer 2011 drilling, and both the mayor and Pete Slaiby turned their attention toward influencing federal decision makers who would ban or allow that to occur, and who would make new rules for offshore oil exploration in light of the *Deepwater Horizon* explosion.

For the mayor, a constant question was when to compromise, when to make demands, how to use his Washington-based lobbyists or legal department. Strategy was complicated by the fact that the region was physically changing so fast sometimes it was tough to know what to do.

Science had become a key part of politics, and that is why a 46-year-old biologist named Brian Person—working for Itta—woke up in a tent on the tundra in late June, 120 miles southeast of Barrow, chomped down a couple of PowerBars for breakfast, bagged a sandwich/apple lunch and with a pilot and a net gunner named Jason Caikoski* (a net gun is a rifle that fires a net), climbed into a small helicopter that had been parked there overnight.

Brian was one of the 24 percent of Barrow residents listed in the last census as white. His PhD came from the University of Alaska in Fairbanks. Originally from Minnesota, the unshaven wildlife

---

* Caikoski was a biologist with the Alaska Department of Fish and Game.

researcher lived with his girlfriend in town and had just bought his first house. That day he wore clothes still wet from the previous day's work: jeans and heavy socks, rubber boots, a raincoat over a polyester jacket over a woolen shirt over a T-shirt. The gloves were cotton, the hat made of otter fur.

The land they flew over was unmarked by human construction, but within ten years might be crossed by oil pipelines coming from offshore. Shell's hope was to connect the new offshore fields to the existing Trans-Alaska Pipeline, which currently ran from Prudhoe Bay south through the borough to the oil terminal in Valdez, 800 miles away. As of that month the pipeline had carried an estimated sixteen billion barrels of oil.

But the area over which Person was flying was also the annual migration route of caribou, the elaborately horned, reindeer-like deer that roved in three major herds across the North Slope, numbering 700,000 in all. Caribou made up an enormous percentage of the Iñupiat diet.

"If whales are the bread, caribou are the butter," Brian liked to say.

That day he was out to check satellite collars on animals in the 64,000-caribou Teshekpuk Lake herd.

"North Slope hunters remove up to 10 percent of that herd annually, up to 7,000 animals a year."

Brian knew that the herd he sought had not yet been exposed too much to oil and gas activities. It provided meat to four North Slope communities and its migration route only skirted Prudhoe Bay. The animals were not used to seeing buildings, roads or pipelines. If new pipelines ended up crossing caribou corridors it would be important, if a spill occurred, to quickly protect this food source.

"We want to know where they congregate in summer so we can stage cleanup equipment in those areas in case they get exposed to oil. Oil on their fur would be extremely detrimental. If there is a spill and we know where they are, we can try to move the animals. And we want to ensure that crucial corridors remain open for migration."

Brian Person spent 40 percent of his time studying caribou. He and other North Slope researchers regularly conducted herd censuses and monitored blood, diseases, diet and illness. That day he was out to service radio collars worn by roughly 50 animals. The collars contained transmitters powered by lithium batteries with a life span of two years. Each collar broadcast at a unique frequency and was linked with a satellite to provide information on the animal's location. It also indicated whether a caribou was alive. The collars contained mercury switches that—if the animal didn't move for 36 hours—would change signals.

The men searched for specific caribou by frequency.

In the front passenger seat Brian looked down at spongy tundra—the green part coming alive in spring, the brown marking dead vegetation from last year. The land was dotted with lakes and cut by freshwater rivers nestled between banks that rose as high as a hundred feet or as low as ten. The copter flew low enough so that Brian could see ripples on the water. Arctic foxes looked up from hummocks, their coats red or tawny brown, no longer winter white. Wolves roved in pairs or alone. Brown bears looked smaller than the big coastal grizzlies or the polar bears farther north, or the odd humped hybrids—part polar bear, part grizzly—that had been showing up as grizzlies moved north with warmer temperatures and the two species mated.

Temperatures had hovered at a brisk 35 degrees Fahrenheit this morning and rose toward the predicted high of 40. Patchy fog alternated with clear areas. Brian's breath frosted.

The cockpit smelled of musty caribou. "Only a hunter or biologist would like it." Bits of wet animal hair blew around with the door open so Jason the net gunner—sitting behind Brian—could have a clear field of fire.

"The project was established by the borough from the very beginning because we knew oil and gas development is coming."

Brian wasn't trying to stop drilling, just prepare for it. He worked for the North Slope Department of Wildlife Management, a small but internationally recognized group of scientists whose findings have played a major role in determining Arctic policies in the United States and abroad.

"Our scientists learn facts that you can use in a courtroom," Itta had told me.

In fact, scientists had been specifically brought in by the borough in the 1970s as a bulwark against outside interference in the Iñupiat way of life.

At the time, oil wasn't the issue. Whaling was. Scientists based elsewhere had estimated the total bowhead population as low as 600 animals. The International Whaling Commission (IWC)—formed to conserve whale stocks and regulate commercial and aboriginal whaling around the world—ruled that all bowhead whaling must stop. Continued hunting of whales even by natives could lead to extinction, the commission felt.

On the North Slope the news was met with shock. The US delegation to the commission lacked a single Eskimo member then, and local hunters insisted that the numbers were far higher based on what they saw. But these claims were viewed as self-interest. The commission relaxed the ban slightly in 1977, but the hunting quota they imposed was still too small to meet local needs.

Rather than going to court or relying on political lobbying, the North Slope Borough brought in scientists.

North Slope and Alaska state researchers camped out on the ice and "came up with groundbreaking work," combining visual sightings and new underwater acoustic techniques to arrive at more accurate estimates, Greg Donovan, chair of the IWC's scientific committee told me.

Donovan added, "Without the financial and moral support of the mayors of the borough, this could not have occurred."

Oops. It turned out there were thousands of bowheads out there, not 600. IWC estimates had been based on an incorrect assumption: that no whales could migrate when the leads were closed, the ocean covered with ice. "This," Donovan told me, "was not in accord with native traditional knowledge."

In the end, native knowledge turned out to be right and the scientists were wrong. The IWC raised the quota.

By 2010, "The ongoing commitment to incorporating traditional knowledge into sound science has been continued by Mayor Itta and has resulted in the bowhead whale hunt being one of the best managed in the world," Donovan said.

The caribou herd that Brian Person monitored that June did not travel in a mass but in strung-out groups numbering 50 to 200. Animals could be spotted from 50 feet up. Brian worked in tandem with other researchers in a fixed-wing plane—with a better view than the copter's—that had balloon tires enabling it to land on tundra. One reason the airplane was there was to rescue the scientists if something went wrong.

The spotter plane circled as the copter located a male caribou they were after. Before getting closer, the pilot landed and Brian hopped out to decrease weight. When the copter rose again, Jason Caikoski stood on the skids, harnessed to the copter for safety, net gun in hand. The pilot moved in quickly, veering, diving, trying to separate out the 200-pound bull male. Steam shot from its mouth as it bolted across ponds and tundra.

"We try to keep it out of lakes, move it to higher ground," Brian said.

The animal zigzagged off and the roaring metallic dragonfly shape above flew so low the rotors sent up spray. The copter came as close as twenty feet from the caribou.

Incredible acrobatics, Brian thought, watching.

Finally, from a quarter mile off he saw Jason fire and the net spread out in the air. A few seconds later he heard the shot. As the copter landed, Jason unhooked himself from the harness and jumped down and Brian was running toward them. The caribou, blanketed in the net, its front legs and head caught in the webbing, was hopping up and down.

"We don't drug animals, because anyone could harvest them within the next 30 days. The drug may not have cleared away by then. We don't want narcotics in the caribou."

Jason charged into the caribou, knocking it on its side. He pulled three hobbles off his belt and bound the caribou's legs to keep the sharp hooves from striking him. All the men in the crew have been kicked at one time or another.

"My left nut swelled up to the size of a kiwi fruit. One of our gunners dislocated his jaw once jumping off the copter," said Brian.

Science on the North Slope is not quiet, lab-bound and theoretical. Researchers disappear into the wild for weeks. That's the way Brian wanted it. He'd loved camping out as a kid in Minnesota. He preferred fieldwork to docile lab work, and that day he was getting a dose of what he liked. When the hobbling was complete, Jason blindfolded the animal to calm it.

"The main purpose of our department is to ensure that subsistence practices can continue as they have in the past, a good use of my education," Brian said.

Brian knew that all over the world, caribou herds are declining, sometimes from illness popping up as weather warms and disease-carrying flies proliferate. Brian's opinion is that climate change is a big factor.

The North Slope herds, however, he said, "seem to be doing well."

The report that would wind its way through borough offices and eventually reach Mayor Itta said that at the moment the caribou seemed generally healthy.

Brian knew that some researchers in the department—on the whale end—were urging Mayor Itta to fight Shell's plans until more science was collected. They wanted proof that development would not harm whales or scare them off with loud sounds. But when Itta had asked the researchers if they had hard evidence showing that the whales would be affected, the answer had been no.

So Itta had not joined in any legal action for the moment. At the EPA an appeal was pending on a ruling that had granted Shell an air-quality permit for drilling. The complaint against the decision had been brought by national environmental groups allied with the Alaska Eskimo Whaling Commission—Harry Brower Jr. was its chairman—and by the North Slope village of Point Hope and a native activist group called REDOIL (Resisting Environmental Destruction of Indigenous Lands). Itta had refused to join the action. He did not trust the way the different sides stopped talking to each other once a legal battle began and decisions became all or nothing instead of compromise.

Nothing that came from Brian Person's caribou study would change Edward's thinking. But he wanted more science. He'd approached Shell to ask if Pete Slaiby would consider funding joint scientific studies of wildlife. Their staffs were negotiating this while Itta remained in a strategic holding pattern, trying to figure out when to cooperate, when to fight.

Then again, there was no movement on the federal level at the moment. Itta and Shell awaited word from the Department of the Interior as to whether the drilling halt in the Arctic would be lifted.

In Anchorage, the pressure was rising for Pete Slaiby to figure out a way to try to speed along the decision.

Some in his office feared losing their jobs.

There was no assurance that the suspension would ever end at all.

CHAPTER 4

Pete Slaiby's Challenge, July

The Eskimo and the oil man could not have come from more different backgrounds. If Edward Itta traced his strengths to a lifelong relationship with one location, Pete Slaiby was a globetrotting ambassador for technology and the only company he'd worked for since graduating college 29 years before.

Royal Dutch Shell was the second biggest energy company in the world in 2010, with overall sales of $369 billion. Headquartered in the Netherlands, its 93,000 employees—twenty times the population of Barrow—generated profits of $18.6 billion in that year. Shell has offices in 90 countries. Roughly 24,000 people work for Shell just in the United States.

"When I see a drilling rig, I see it as aesthetically pleasing," Slaiby told me.

Shell was also widely considered a pioneer in offshore drilling. The company opened up the Gulf of Mexico in the 1960s by unveiling a new floating drill platform, a development so revolutionary, wrote industry analyst Tyler Priest, "it was akin to John Glenn's space orbit." It made drilling possible in waters greater than 300 feet deep. Then in 1975 the company's new seismic technology pinpointed oil in the Gulf below 1,000 feet. To reach it, Shell constructed the first deepwater drill platform, winning the 1980 award for outstanding civil engineering achievement from the American Society of Civil Engineers.

No oil company had ever won that award before.

"When I see a drill rig, I see the shape of things coming together."

The future oil executive came from Manchester, Connecticut, a Hartford suburb, and from an orderly block where every house stood 90 feet back from the street, every home sat on a three-quarter-acre plot. Life was math. His parents would still live in that house 50 years later.

Pete Slaiby's dad, Ted, worked as an aeronautical engineer at Pratt & Whitney for 37 years. Mom taught elementary school. They owned a small farm that the family worked on weekends. Asked to tell a story characterizing his father, Pete selected one about Pratt & Whitney engineers' refusing to quit work on a jet engine they believed in, risking the company's solvency in a bid to win a contract. Nobody outside the company thought they had a chance, Pete said. They won it.

"Hard work is good for the soul," Ted told Pete.

Pete's brother would become a vascular surgeon. His sister would join the Peace Corps and get a master's degree in environmental policy.

Like Itta's parents, Slaiby's valued education. But the opportunities in Connecticut were more mainstream America, the kind arising in a place filled with white collar jobs, well-trimmed lawns and expectations that kids attend college. Flagging grades meant punishment.

"In his early teens Pete started paying more attention to his buddies and his grades dropped," Ted Slaiby said. "We sent him to a Catholic school for a year. It was rigorous. Those days, discipline was really followed in schools. Pete admitted he'd been goofing off, said he'd learned his lesson. After that his grades were good."

Pete played Little League in junior high and swam on his high school team. "There was some underage drinking and getting thrown out of bars," he said with embarrassment but also pride. He bought a Mustang convertible with money earned from an after-school job delivering flowers and another teaching swimming. July days found Pete in the lifeguard chair at a concrete-lined reservoir, scanning kids splashing below. He liked cars, partying and work.

"I decided I wanted to be a biomedical engineer."

Senior year in high school brought the happy acceptance letter from Vanderbilt University—Harvard of the South—and at the point in life when Edward Itta entered Native American technical school in Cleveland, Pete Slaiby arrived at his Nashville, Tennessee, dorm that fall to help carry boxes up the stairs for another kid moving in. Dave Moore of Arkansas would become a lifelong friend.

"Pete had a little bit of a nerdy air, like Ben Stiller in the movie *There's Something About Mary*," Dave recalled.

Responsible. That's how Dave saw Pete. "If we went out on a Saturday night, he'd be in church Sunday morning. We were amazed. First time away from home, but he'd go every week."

Considerate. Asked if he'd ever seen Pete scared of something, Dave said, "There was this girl who liked him. Pete liked her just as a friend, but he was afraid he'd hurt her feelings so he went out with her a few times anyway."

Ever since that Catholic school experience, Slaiby had priorities when it came to education.

"The Mardi Gras was on in New Orleans and Pete said he was going and would be wild and crazy. He had tests coming up. We didn't believe him. Well! He goes upstairs and packs and took off in the middle of the night. The next morning he was back. He'd been driving in Alabama when his responsibility kicked in. He turned around."

Pete switched majors to mechanical engineering. The B+ student enjoyed road trips to football games and drinking beer at the age when Edward Itta was in the Navy. Slaiby had a long fuse and a good sense of humor, the guys said. He wasn't afraid to make fun of himself. "When he drank he became Superman," Dave recalled. Drunk once, Pete jumped off a second-floor bar balcony and hurt his leg. After that the guys nicknamed him "The Wounded Caribou."

He had not aimed to work in the oil industry, but oil supply was running low in the United States by the time he got his bachelor's degree. "The oil crunch formed my career. Shell brought me down to New Orleans for an interview and made me an offer as I was leaving. I liked that decisiveness."

He started out in the New Orleans office as a petrophysical field engineer, working out in the Gulf of Mexico, "evaluating wells, getting lots of experience."

By his mid-twenties he was overseas in Syria, working for a Shell subsidiary, Pecten International, helping build an oil-processing facility in the desert. Like many Shell employees he was beginning years of moving around—working with other cultures. He liked it.

"It was fun to visit the Bedouins and see how the project fit in. I lived in Damascus across from the Badr Mosque. It was gorgeous. Weekends we'd have runs through the old part of the city, then sit around and drink."

After Syria came Brazil, where Slaiby supervised work offshore in the deep waters of the Santos Basin, on that country's continental shelf, on a gas project jointly owned by the corporation and Brazil's national energy company PetroBras. He played lots of tennis. The old college pals were amazed at stories he told during what would become annual reunions in Nashville. Pete had a full-time driver in Brazil. He was globetrotting and didn't seem so semi-nerdy anymore, Dave thought.

"I spent a lot of time on the beach and probably got melanoma in Brazil," Pete joked.

He also met his future wife. Rejani Machado was a pretty, good-hearted, family-oriented Brazilian who fell for Slaiby the night she met him when a group of young people went to a Tears for Fears pop rock concert in Rio.

"He was the boyfriend I had always wanted."

She thought the American boy kind—he bought chocolates for the women—and good looking, and she asked Pete's Portuguese teacher if he had a girlfriend. The wry reply was, "Lots." Rejani found out which Ipanema beach Pete frequented and began showing up there, taking along her sisters and mother, who brought home-cooked food.

"My mother believes you get the man through the stomach." Rejani laughed.

After they started dating and Pete didn't propose, Rejani took a job in Italy. Pete visited and invited her to Connecticut for Christmas. The return ticket, she noticed happily, took her back to Brazil. No more Italy.

"We honeymooned in South Africa."

The globetrotting couple spent a pleasing four and a half years in England, where Slaiby worked offshore on North Sea gas wells; four and a half years in Cameroon, Pete offshore again; then Brunei, where Shell had platforms in the South China Sea; and back to Brazil, Pete's favorite.

"The work always involved partnerships with local people," he said, meaning that there were always unique problems that had solutions and, in his memory at least, locals were always satisfied with the result.

"They were improving their country and their lives."

The life was intoxicating. In England, if Pete was away, Rejani might take a train to Holland for the day just to get her hair done. "But I'd be back by the time Peter got home." Cameroon? Brunei? "We loved all the places."

Pete was building a record as someone comfortable with living in a variety of cultures and working well with people there. This was noticed in Houston.

His parents and old college friends were always asking if the foreign assignments were safe. Slaiby *loved* them. "You get on a roll with this expat thing. The opportunity to live beyond being a tourist, to understand another culture. The money is good and goes far overseas. The lifestyle is addictive. You're challenged to assimilate, become effective. That becomes an end in itself."

If favorite books reflect personality, Pete's top pick remains *On the Road*, the 1957 Jack Kerouac story about beat generation friends traveling around the United States. Pete bought many copies over the years and regularly brought one on trips. He described his favorite character in the book, Sal Paradise, as "a seeker who continually finds a way to enjoy life."

"Is Sal Paradise you?" a reporter once asked him.

"Maybe a more mature version."

Finally in 2008, Dave Lawrence—Shell's executive vice president for exploration, asked Pete if he wanted to move to Alaska.

"I said no, what else do you have?"

Dave was the exec who had decided that the Alaska "play," as oil prospects are called, was huge. He had led the appraisal resulting in Shell's record-breaking bids. Alaska was Dave's baby and he suggested the move to Pete again a few weeks later.

"Can we look around for something else?" Pete said.

Dave is a low-key guy. Three weeks went by. "Hey, how about Alaska?" he said.

"You learn to read the tea leaves." Slaiby sighed.

He *did* have a choice though. He could go to Alaska or Egypt. Pete and Rejani flew to Anchorage and were impressed by the beautiful city on Alaska's southern coast. The neighborhoods looked good and the schools too, as they hoped for a child. Rejani thought she could make the move.

The Alaska situation—Shell's grand vision—had mired down into a mess. Someone was needed who could deal with roadblocks thrown up by the Eskimos, lawsuits and bumbling federal agencies.

"I was parachuted in."

In the end, of all the places Slaiby worked over the years— third-world countries, jungle countries, a "dictatorship" renowned for being difficult—Alaska would turn out to be the hardest, he'd tell people.

"Difficulty level? Nine out of ten."

The problem started with government regulations. More and more had been added piecemeal over the years. Their goals were noble—to protect the environment—but neither Congress nor the White House under either Democrats or Republicans had done anything to coordinate them.

In 2008—when Pete arrived in Anchorage—any proposed drilling project in federal waters off Alaska had to satisfy requirements detailed in laws including the Outer Continental Shelf Lands Act (rules for lease sales and drilling), Endangered Species Act (don't harm animals), Marine Mammal Protection Act (protect marine mammals), Magnuson-Stevens Fishery Conservation and Management Act (protect fish), National Historic Preservation Act (protect historic sites), Clean Water Act (protect water), Coastal Zone Management Act (do only what states allow if they have a coastal management plan) and the Clean Air Act. The granddaddy law, the National Environmental Policy Act, required that any construction slated for federal lands go through an environmental analysis before being approved.

There was no requirement for all overseeing agencies to coordinate studies, decisions and permit giving.

The rules kicked in before a lease could be sold. The Department of the Interior's Minerals Management Service (MMS), for instance, had to evaluate the environmental impact of any proposed oil or gas lease sale before giving it the OK. That called for consulting with the US Fish and Wildlife Service and NOAA's National Marine Fisheries Service to see if the sale might damage a commercial fish habitat, endangered species or marine mammals even if they were not listed as threatened by the federal government.

Once the analysis was done and the lease sold, the company buying it could still not drill yet.

First the company had to submit an exploration plan to be evaluated by, again, the MMS. This would detail *how* the company planned to get at the oil.

The logic was paying for a lease didn't give a company license to tear apart the sea bottom in a reckless hunt for oil. You had to show you would act responsibly.

If MMS approved the plan, that *still* did not entitle drilling, though. At least 30 federal permits were required before a company could do that, and *getting* the permits called for more analysis and public hearings.

For instance, if Shell decided to build an onshore facility and a road crossing—say, a river on federal land—Shell needed permission from the Corps of Engineers. If they wanted to bring in a drillship to look for oil, they needed an air-emissions permit from the EPA.

After the EPA awarded the permit, the decision would undergo a review inside the agency by an administrative board if any citizen or group complained. The board could negate the permit. The initial application process could take months. So could the review.

In 2007, for instance, an EPA environmental appeals board sent Shell's permit back for reworking after Edward Itta's North Slope lawyers filed an appeal.

And even if the clean-air permit went through in the end, that wasn't enough. Anytime boats were to be operated in Arctic waters— seismic boats, drill rigs, survey ships—a permit was required from NOAA's Fisheries Service outlining rules by which the crew must stop work or alter course if they spotted whales, seals or walrus ahead. If the Fisheries Service okayed the permit, the recommendation would be reviewed by higher-ups at NOAA.

Permit applications were not short documents like automobile license requests. Just the application for the clean-air permit could run 1,400 pages, weigh seven pounds and commonly take nine staffers months to compose.

Also, some permits had to be renewed annually. So if the process got held up one year, Shell staffers had to start from scratch the next.

The whole system would have been complicated enough had the MMS been competent, but by 2008 other government agencies reported that not always to be the case.

A Government Accountability Office (GAO) report on offshore oil and gas development charged the MMS with suffering from destructive "high staff turnover" and said that employees reported "lack of guidance" to do their jobs right.

The Outer Continental Shelf Oversight Board charged the MMS with lacking "a formal bureau-wide compilation of rules, regulations, policies or practices pertinent to inspections."

Then it got worse. MMS was charged by two of their own scientists with suppressing or changing research results so as to favor drilling plans, according to the GAO.

In June 2010, following up on accusations by the Washington,

DC, advocacy group Public Employees for Environmental Responsibility, *Denver Post* investigative reporters found old e-mails confirming that Alaskan MMS officials—alarmed by warnings by one of their scientists that a large oil spill would probably cause "significant adverse effects" on fish populations—had during the George W. Bush administration left a handwritten note on the scientist's desk saying that this conclusion, if released, would "delay a 2007 lease sale in the Beaufort Sea. That as you can imagine, would not go over well with headquarters and with others."

When the scientist refused to change the findings, a manager in the office did.

Another MMS scientist concluded that Shell's exploration plan did not take the potential effect on polar bears into account and got an e-mail back from management saying, "Now we're in an extreme time crunch and under intense pressure to get the environmental assessment done regardless."

A big reason for the suppression of research, critics felt, was that a fundamental conflict of interest existed inside MMS. The agency had two opposite roles. As the entity charged with analyzing drill plans, it was supposed to protect offshore environment. As the agency charged with raising revenue for the government through lease sales, it was encouraged to approve plans fast.

These flaws were food for lawsuits. Drilling opponents were always looking for ways to stop offshore Arctic exploration, fearing it would lead to catastrophic spills. Lawyers from organizations including the Sierra Club, Alaska Wilderness League and Oceana constantly sought ammunition.

"The laws are there for a good reason," said Peter Van Tuyn, an attorney whose clients included the Alaska Wilderness League and the Native Village of Point Hope.

It was Edward Itta who had come up with the lawsuit idea in the first place. "The borough led the way," recalled Jim Ayers of Oceana.

"Mayor Itta was directly responsible, under a tremendous amount of pressure to stay out of things. He spearheaded the push to have the federal government do science first before they allowed drilling."

In 2008, Itta's suit—also brought by the Alaska Eskimo Whaling Commission and several native and environmental groups—resulted in the Ninth Circuit Court of Appeals voiding Shell's exploration plan in the Beaufort Sea. The ruling said the MMS environmental analysis had gaps in it so big that its "abrupt conclusion" that any potential negative impact of drilling would be insignificant was "unsubstantiated."

Even if Itta didn't sue, he had a bully pulpit to speak for the native inhabitants of the North Slope.

In short, by 2010 it was Itta's approval, condemnation or silence that Shell *and* environmental groups sought as they maneuvered to influence federal decisions.

Curtis Smith, Shell's Alaska spokesman, said, "If the North Slope is against you, you won't get anywhere."

Pete Slaiby said, "All we want are rules that are clear and the knowledge that if you follow them you'll achieve a desired outcome."

Pete was finally back on American soil but *still* considered himself an expatriate. "Houston is 4,000 miles from Anchorage. We're a remote office. Alaska is different from the lower 48, a whole different country and part of the same country at the same time."

His staff was filled with other expats. Susan Childs, responsible for permitting issues, came from Corpus Christi, Texas. Trained as a biologist, she'd once worked for the Minerals Management Service but preferred Shell, where hard work was rewarded, she said, to government service, where, she said, it was not. Her husband, Jeff, had worked as a biologist, also for the Minerals Management Service.

"There's blood on the walls at home," he joked.

Phil Dyer, an "issues and stakeholder manager," spent a lot of time in North Slope villages. His job was to spot political problems before they came up, and to prepare the company on how to deal with them.

He came from Great Britain.

Peter Scott oversaw the communication division and had recently moved to Anchorage with his wife and two young daughters from Australia. Dr. Michael Macrander, Shell's head of scientific research in America's Arctic, was an ex–university professor from Alabama. Mark Duplantis, wells delivery manager—in charge of actual drilling—hailed from Louisiana. The Shell people tended to work together, spend weekends skiing with one another and dine often at one another's homes, reinforcing internal points of view.

Pete Slaiby's two top Alaskans on staff were Cam Toohey, a political expert who'd formerly worked for the Department of the Interior, and Curtis Smith, a popular Anchorage ex–TV anchorman, now press spokesman.

Up on the tenth floor of Anchorage's Frontier Building, with its fine views of the Chugach Mountains and the Minerals Management Service office, Shell executives tried to figure out how best to explain to the public the difference between an *exploratory* well, designed to locate oil, and a *producing* well, which sucks it from the ground.

They were sure that if the public understood the difference, much opposition would die away.

Shell was only asking to drill simpler exploratory wells. The job would take a few short summer weeks, then the drillship would leave, the well would be capped, and by the time the sea froze over nothing would be left to see. There would be no big *Deepwater Horizon*–type rig on the water.

Only *if* the exploratory wells struck oil would a drill platform be built. But before that could happen a whole new set of federal regulations—more safeguards—would kick in and hundreds of new permissions would be required.

"We don't even know if we'll find anything down there in the end," Pete said with a sigh.

Slaiby knew that in the best of all possible worlds, *if* the exploratory wells hit oil, between the time required to obtain new federal, state and borough permits after that; construct a production platform able to withstand 30 feet of moving sea ice; and also build a pipeline to shore, a minimum of ten years would pass before any actual oil or gas came out of the ground.

No wonder the situation was a mess.

The Iñupiat people had lived on the North Slope for 4,000 years. Most Shell people who flew up to meet with them had lived in Alaska for scarcely four.

Told that North Slopers often resented it when he spoke about his former postings for the company, as if there were a similarity between Brunei and Barrow, Pete didn't back down. "I can believe it. But people in Brunei love their environment every bit as much as people in the North Slope love theirs. There's no exclusivity in wanting to protect where you come from. In Brunei we worked in the South China Sea, in live coral areas and mangrove swamps. Those lovely areas required the same amount of caution and care you need to operate with anywhere in the world."

At the same time, he knew that each job came with unique problems. Slaiby's library at home was filled with books on Alaska's Arctic. He knew about the Yankee whalers and the diseases that decimated the Iñupiat, and about dangerous plans that outsiders have tried to push through before on the North Slope.

He knew how resentments still lingered over the worst of these, Project Chariot, which shocked North Slopers when they first heard about it on July 14, 1958, the day that Edward Teller, father of America's hydrogen bomb, flew into Juneau, Alaska, to announce a plan to create an instant deepwater harbor beside the village of Point Hope by blowing up several thermonuclear bombs there.

"We have learned to use these powers with safety," Teller assured listeners.

Villagers would leave temporarily and happily return after the event, he said. The blast would pack firepower 160 times larger than Hiroshima. Teller said that the Atomic Energy Commission, project sponsor, could control the explosion so perfectly that it could "dig a harbor in the shape of a polar bear if required," Slaiby read in Dan O'Neill's superb book *The Firecracker Boys*.

With the enthusiastic support of Alaskan newspapers, politicians and academics, workers were hired. Bulldozers arrived at the village. Scientists decided to test how radiation from a blast might spread in local water, so they sprinkled soil contaminated with cesium 137 and strontium 85—its radioactivity a thousand times stronger than federal law considered safe—around the area.

Meanwhile, Atomic Energy Commission spokesmen—unaware that the Eskimos had brought tape recorders—told an audience of Point Hope villagers that radioactive fallout from any blast would be so little it would probably not even be detectable and that Japanese survivors of the atomic bomb blast at Hiroshima, having received "very great exposures" recovered and later experienced no further effects, Pete read in O'Neill's book.

Eskimo opposition was credited with halting the project, but even so, raised radiation levels had been found in Point Hope in spots as late as the 1990s. It turned out that some contaminated soil had never been removed.

"Project Chariot put a huge context on any meeting I had in Point Hope," Pete said. "When they asked if we had any technology to detect radioactivity, I saw that they associated any kind of development with Project Chariot."

Slaiby also knew about Shell's own blunders in the Arctic. Shell had not helped its own cause when it returned to Alaska in 2005 after seven years away. The company had drilled 32 offshore wells there earlier without accident, so Houston assumed they'd be welcomed back.

But North Slopers remembered things differently. Whale hunters said that company activity had deflected whales then, driven them farther offshore, making hunting them more dangerous for men in small boats.

Rather than reacting to these concerns, Shell officials informed villagers of their plans but would not change them.

As Rick Fox, Slaiby's predecessor, said, "We felt that even if local interests wouldn't come along with us, the federal government had sold us leases so they'd give us the permits."

Instead, Shell execs were blindsided. Villagers demanded to know how Shell could clean up any spilled oil. They wanted promises that ships would stay away during hunting season. They feared that oil spilled during winter under ice might not even be *detected* until spring. Rick Fox seemed to many to be evasive on these questions at community meetings. *That's not what we're talking about right now.* And bad feelings were compounded when Rick hired former North Slope mayor George Ahmaogak (pronounced Ah-ma-o-yak) as a Shell consultant after Edward Itta defeated George in the mayoral election.

"We had not seen this kind of resistance before," Rick told me.

To many on the Slope it appeared as if Rick was trying to buy influence, not listen to concerns. And that Shell lacked understanding of how power worked in Eskimo communities. It wasn't like corporate decision making. There was no single leader. The North Slope was divided into groups that included village tribal associations, local or regional whaling captains, and native-owned corporations, and the same people belonged to different groups. Decisions were reached after much discussion. Outsiders were expected to talk with all groups, not hire one person whom they thought could influence them all.

By the time Pete Slaiby took over, resentment was high, distrust rampant, and he knew that to many he looked like the latest in a long line of outsiders who appeared out of nowhere, promised benefits from some plan and tried to push it through whether Eskimos liked it or not.

"They are a very sophisticated audience. They've been around oil for years," Slaiby said.

"Also," he added, "it turned out that there was one year where our work did deflect whales off one village. It was the seismic boats, not the drilling."

Slaiby had made some headway by 2010 in the sense that, as Itta aide Andy Mack said, "Pete Slaiby is not a smooth talker. He's a blunt guy. He knows he's going to walk out of a meeting with a black eye. But he says what he thinks."

Slaiby had made less progress with federal agencies that worked out of offices far from Alaska and were unfamiliar, he felt, with conditions. EPA regulators were located in Seattle, 1,400 miles away. Sometimes foot-dragging bureaucrats didn't seem to appreciate that the Arctic offshore exploratory drill season is only three months long. Drilling is conducted from a ship that must depart when autumn ice moves in. That means if an agency delays a decision, an oil company can lose a whole season's work.

By 2008 no exploratory well had been drilled in federal Arctic waters in eight years, although leases had been awarded to companies including Shell; ConocoPhillips; Total, from France; and Statoil, the Norwegian national oil company.

By June 2010 nothing had changed.

Now as summer approached, even though actual drilling was out of the picture that year Shell needed to prepare if it was to continue its hoped-for program the following year.

Shell had hired helicopter crews and an archeologist to survey shore areas where a pipeline might come ashore. They needed to locate historical sites like centuries-old abandoned villages that, under law, the pipeline must avoid. Shell hydrologists needed to survey coastal watersheds. A pipeline would have to steer clear of places where drinking water or animal habitat could be affected by construction or a spill.

Shell had also leased boats to conduct seismic activity offshore called shallow hazard surveying. This involved mapping the sea bottom where pipelines might run to land. The purpose was to look for dangers that might damage a pipeline, such as sunken ships, or to find gouge marks from icebergs. Any pipeline would need to be situated below gouge marks so icebergs wouldn't damage it.

But in order for the surveys to proceed, Shell required an "incident harassment authorization" from NOAA's Fisheries Service that would outline shutdown or avoidance procedures if boats got too close to marine mammals. The permit was being held up by higher-ups at NOAA, Shell executives believed. It was certainly taking longer to get than usual.

Shell's boats were sitting in Dutch Harbor, in the Aleutian Islands, eating up thousands of dollars in costs while the company awaited a decision.

For weeks, phone calls to the agency went unanswered. Even Shell's chief lobbyist in Washington, Brian Malnak, was having trouble getting through.

Slaiby was frustrated because Shell wasn't even asking to drill, just to operate seismic ships, which they had done for years.

The application kept sitting on desks. NOAA officials would claim later that many authorizations were held up during those months as the Fisheries Service took extra care analyzing requests in the wake of the Gulf disaster, that no animus had been directed at Shell or any other oil company. And that decisions were made in a timely manner. But even officers on a Coast Guard icebreaker conducting research off the North Slope in 2010 found that harassment permissions were late in coming from Washington that year.

Either way, another regulation was stopping Shell.

And even if the company got the authorization for the ships, Secretary of the Interior Ken Salazar had not announced whether the Department of the Interior drilling suspension in the Arctic would end by the 2011 open-water season. He was still making up his mind.

"We had to figure out the art of the possible in 2011," Cam Toohey said.

The Deepwater Horizon Commission was scheduled to review Arctic drilling plans, but their report was not expected out until January. Shell's choice was to keep pumping money into the 2011 venture or call it quits again.

In all, by this time the company had sunk $3.5 billion into the Arctic project.

The fear at Shell was that the *Deepwater Horizon* incident would halt drilling altogether. After the *Exxon Valdez* oil tanker ran aground in March 1989 on Bligh Reef in Alaska, "it took three years to get back to normal," said Slaiby.

"Our big question was whether the Obama administration still supported drilling at all."

Bottom line? By summer 2010 the absurdly complex federal process was driving Pete Slaiby and Edward Itta both crazy, keeping them in the dark for months while decisions were delayed. They both felt at the mercy of the leviathan thousands of miles away, although they sought different results. Slaiby wanted offshore drilling, Itta more guarantees on environmental protection but also assurances that some oil would flow south from the North Slope.

Their mutual unhappiness with the process was so marked that occasionally Slaiby's or Itta's staffers wondered in private if there might come a day when, driven by frustration, the men would actually stand up together or separately to speak up in favor of *some* Shell plan. After all, the thinking went, Shell kept shrinking its plans in response to Iñupiat suggestions and the mayor kept talking about the need for *some* drilling on the Slope.

These thoughts of cooperation were battered away.

Oil supply was not the only important Arctic issue facing the country as that summer began and not the only one to play out near Barrow. In July a long red icebreaking ship eased its way north through the Bering Strait—the narrow passage separating the United States and Russia—and headed for a rendezvous with a Canadian icebreaker coming from the east. On board were scientists whose work over the next five weeks could help add thousands of square miles to US-owned territory north of Barrow and eventually trillions of dollars to government coffers and national businesses.

The icebreaker was Coast Guard's *Healy*.

Results of the mission were to go to the White House and State Department but kept away from the press.

The mission was part of a land rush in the Arctic, the biggest distribution of lands on earth ever to occur at one time, and a race in which the United States lagged behind.

CHAPTER 5

Aboard Coast Guard Icebreaker *Healy*, August

The mountain appeared as spectacularly as a volcano rising from the South Pacific. Its sheer escarpments loomed above a flat plain extending in all directions for hundreds of miles. The diamond-shaped base rose to a crest 3,300 feet tall.

The mountain had no name, though.

No human had ever seen it.

It sat 11,000 feet beneath the Arctic Ocean, covered by ice as impenetrable as clouds obscuring a tropical isle from passengers overhead in a Delta jet.

On August 25 the big red icebreaker bulled into ice above the mountain, backing and ramming through heavier spots, nudging aside small floes, its route guided by satellites in space and human eyes on the Aloft Conn—a lookout reachable by climbing up three indoor ladders from the bridge—100 feet above the sea. The Coast Guard cutter *Healy* had reached latitude 81.34 North, 700 miles north of Barrow, and west of the Canadian Arctic archipelago.

We were in international waters 500 miles from the North Pole, and the ship's mission was to provide the State Department with evidence to help extend US sovereignty over an area of Arctic sea bottom the size of California. Perhaps the mountain or nearby areas would not *remain* international. Perhaps—because of a new treaty—extra territory might soon belong to the United States.

Elephant country. That's what oil geologists call potentially rich areas. We might be in elephant country.

And this particular morning's job was to scoop up a sample of mountain and bring back mud that might help confirm a US claim to the oil resources and undersea minerals in far-north elephant country.

At 6:30 a.m. the engines slowed to one and a half knots, same speed the ice was moving. The ship drifted in the current as the officers figured out how to position the *Healy* safely during the upcoming *coring* job.

"That's because we can't break ice and core at the same time," said Chris Skapin, Lt. JG, officer of the day, standing watch on the bridge.

The *Healy* is 420 feet long and can carry up to 50 scientists in addition to its crew. It has a wide appearance and with 4,200 square feet of lab space functions mostly as a floating research platform. It offers working decks, cranes, sample freezers and data-transmission ability to scientists.

In open water the *Healy* can move at seventeen knots but when encountering ice up to four and a half feet thick, it slows to three knots to break through. Ice is pushed away slowly in masses. Fault lines open like crevasses during an earthquake. If ice thickens further, up to eight feet, the *Healy* can back up and ram through as long as temperatures remain above 50 below zero. After that—if ice grows thicker or cold more extreme—the cold locks up the engines. The ice locks in the hull. That's why the *Healy* never goes north in winter.

Leaving from its home port of Seattle, the ship had passed

through the Bering Strait and entered the Arctic in early August, passed within fourteen miles of Shell's Chukchi Sea leases, and started a zigzaggy path that—when mapped—resembled a route taken by a drunk riding a lawn mower. But the direction had been determined by the State Department, directing research by scientists to fill in gaps in knowledge about the makeup of the sea bottom. This was the sixth in a series of annual mapping cruises crucial to US interests. At first we had seen no ice even in areas that—according to books in the ship's library—had once been so thick with it at the same time of year that the ice had regularly imprisoned and crushed explorer or whaling ships. This happened well into the 1900s.

Reading books like *The Ice Master* by Jennifer Niven while on the bridge was quite an experience because you could see photos or lithographs of towering ice mountains gripping the old ships. The date on the lithographs would be a century ago. But you'd look up and see open water. Not a floe in sight now.

The sky was generally gray in summer 2010 and the ice, when we'd finally encountered it, had been membrane thin, a glaze of interlocking fingers. Farther north it had gone slushy, then harder, but never had it resembled the paintings in the books. There had been wildlife sightings; an occasional pod of beluga whales. A lone polar bear looking back from solid ice, appearing yellowish from fat beneath its fur. The doglike heads of spotted seals popped up. They were taking quick looks at the ship.

August 25 was misty. Solid daylight had begun at 3:20 a.m., although in summer the sun never went all the way down. From the bridge an ice planet stretched away, dusted with fresh snow. No ships or animals were visible that morning. A lone set of arctic fox tracks angled off in front of the prow. The temperature outside was a summery 27 degrees, warm enough for me to take a mug of bridge-brewed hazelnut coffee outside and breathe brisk polar air as I watched one scientist, the daily exercise walker, march around a lower deck.

Soon crew members and scientists assembled on the bridge for the briefing on the day's upcoming dangers. Crew wore bulky hooded sweatshirts that read HEALY CREW in gold, and red knit caps given to "polar bears"—those who had passed through an Arctic hazing ceremony. Others preferred blue zip-ups or peaked caps. Scientists and technicians were comfortably dressed in jeans, sweaters and fleece pullovers. The North Slope's native community liaison aboard, six-foot-one whaling harpooner Ralph Kaleak, a quiet man, wore a T-shirt that said ORANGE COUNTY CHOPPERS and took notes that would reach Mayor Itta.

"I was told to write down coordinates and watch all the tests. I was told to see if the water temperature is changing. I was told to help," Ralph said.

A master chief (noncommissioned officer) who would help direct operations reached up and- boots on deck—casually rocked back and forth by holding on to a wooden bar crossing the ceiling. It was there to provide balance in case the ship hit rough seas. But in ice, seas were usually calm.

From the bridge, wraparound windows provided a 360-degree view and instruments enhanced it. There were no fixed landmarks, just changing ice. On radar the ice showed up in yellow. Two sets of monitors showed route, depth, ship position. Erin Clark, a wry, funny, ice services specialist from the Canadian Ice Service, stood with her handheld computer, tuned to satellite observations as she helped to forecast open routes, avoid heavy floes. Six a.m. was also a favorite time with scientists or civilian riders—such as the two high school science teachers who'd won a competition to be aboard—who liked to look for Arctic wildlife. I joined a group clustered around 49-year-old Capt. William Rall, for whom this trip was a fifth mission in the Arctic.

Today's "piston coring" would involve lowering by cable, from a high winch above the aft deck, a long steel pipe or "core barrel" shielding a plastic liner, or tube inside. A 2,300-pound weight would speed descent. If all went well the pipe would penetrate the mountaintop in the cold dark below. At impact, interlocking steel fingers inside would be pushed open by incoming sediment. As the pipe burrowed lower, sediment would stream into the opening. When the pipe was winched back up, the mud—trying to slide out—would push closed the steel fingers, trapping the sample inside the plastic tube.

"Stick it in the bottom. Get it out as quickly as possible," said Dale Chayes, science systems engineer.

If things went well today, the sample obtained would join a vast flood of data submitted by the State Department eventually to a UN-based committee in New York. *Healy* scientists hoped to prove to the committee that the seabed north of Barrow was an extension of the US continental shelf, that the makeup, age and order of sediments on land under Barrow's caribou hunters were the same as ones hundreds of miles out. That over millions of years, like dirt running off a tilted lawn, the sediments had washed into the sea to help form the continental slope.

If they could do that, *prove it was the same sediment and formation*, the committee would hopefully decide that the data justified a US claim, under the Convention on the Law of the Sea.

Lots could go wrong though today, and the crew had to be ready if it did. The steel cable, snapping, could decapitate someone or sever a leg as it whipped around. Heavy ice, drifting into the cable, could snag it. What would happen if the winch broke? Or there was a loss of control? Or if the ship lost heading?

Listeners paid close attention. We knew from the commemorative plaque on the mess deck that even routine tasks in the Arctic can turn deadly. A tragic example occurred in 2006 when *Healy* divers Lt. Jessica Hill and Boatswain's Mate 2nd class Steven Duque decided to go for a standard training dive, 500 miles north of Barrow while a party was in progress on the ice. This is called "ice liberty." Crews love it. The ship stops. The sailors, in shifts, go down onto solid ice for a break. Guys take off shirts and snap photos. Girls spread blankets and lie down in bikinis for a minute and run back to the ship. Scientists and crew play touch football. On that day some were drinking beer.

Both divers were connected by tender lines to crew members who were supposed to feed the lines out slowly as the divers descended, and stop if they felt a single tug on a line. But the lines played out fast and inexperienced watchers misinterpreted the signals. The divers had loaded on too much weight. By the time they were hauled to the surface they were dead.

So even on a ship where safety drills were practiced constantly—fire drills, evacuation drills, flooding drills, smoke drills—the coring called for precautions. Scientists and tech people were barred from the aft deck unless they wore hard hats, steel-toed boots and bright orange zip-up float suits designed to keep them buoyant and warm—for a few minutes—if they fell in.

Spaced around the railing during coring, crew members would stand like guards wielding ice gaffs—twenty-foot-long barbed poles. They would repel ice drifting toward the cable once it entered the sea.

Shortly after breakfast, chief scientist Brian Edwards of the US Geological Survey gave the ready-on-deck signal and the coring pipe was lifted off deck and lowered into the black water where the *Healy* had cleared away ice.

Brian was bald, mustached and cherubic faced—a self-effacing, diligent leader who always answered "Best day of my life!" if asked how the day was going, no matter what problems loomed. Moving between the coring and a glassed-in science conning station overlooking the deck, Edwards paid close attention. So did Captain Rall and wisecracking tech guru Dale Chayes from Columbia University's Lamont-Doherty labs. Also present was Pablo Clemente-Colón, probably the foremost sea ice expert in the United States—a tall, cigar-loving Puerto Rican, chief scientist at the National/Naval Ice Center in Suitland, Maryland. Asked how a Puerto Rican had gotten interested in ice, he joked about liking ice in his rum and cokes. Pablo spent hours on his laptop computer each day on the bridge, comparing satellite ice images with what he saw in person. Satellites show ice cover but not thickness, and it is the diminishing thickness, he said, that is crucial to understanding the extent of ice loss in the north.

As the piston corer dropped lower the intercom announced progress as in a NASA spaceflight control room.

"Ice on starboard!" Below, on deck, crew members at the railing kept the floating bits away with gaffes.

"We're at 2500!"

"Make our slowdown depth 2800 meters," said Dale Chayes.

"If we get to 3200 and we haven't tripped [hit bottom] then the ship is drifting," Dale said.

Gathering the evidence was painstaking work in remote areas scattered across thousands of square miles.

Andy Stevenson, 60, an Oregon-based geologist aboard, put some perspective on just this one day's coring process. "Imagine standing on top of the Empire State Building. It's midnight with no moon. You have a thimble with a long piece of thread and you let it down to the New York pavement and scoop up one thimbleful, bring it up, and from that you try to explain the entire origin and evolution of New York City. Geologists live on inference and intuition."

The coring pipe hit the mountaintop at 10:46 a.m.

But was the coring working? No one knew yet.

Light snow fell on the crew on the aft deck.

President Harry Truman started the race in 1945. The Truman Proclamation came just after World War II when it was suspected that substantial oil discoveries might be found in the Gulf of Mexico below the US continental shelf. That's the long shallow slope extending from land into the sea for miles before dropping off when the continent ends.

Truman, "aware of the long-range, worldwide need for new sources of petroleum and other minerals," announced that the US from that day on had the exclusive right to explore and acquire mineral and fishing rights on the shelf beyond what until then had been the recognized limit of any country's national territory—three miles.

Other oceanfront countries liked Truman's idea, so soon Chile and Ecuador announced that *they* would extend their undersea dominions too. Ecuador claimed all seas 200 miles out, including rich fishing grounds for the US tuna fleet.

By the 1970s Ecuador was regularly seizing San Diego–based tuna boats, saying that they violated sovereign waters. More countries announced expansions of their undersea territory. This did not please big powers like the United States or the Soviet Union. Both feared that if claims weren't kept in check, their Navies might lose freedom of navigation—the ability to go where they wanted without asking permission—in the world's key transit straits. There were more than a hundred of these critical bottlenecks around the world, like the Strait of Hormuz connecting the Persian Gulf with the Gulf of Oman, or the Strait of Dover between England and France.

Suddenly the worry was, could a nation actually claim strategic shipping lanes as national territory and try to limit transit, charge entry fees, demand identification from Navy vessels trying to pass through?

Since much naval strategy involves the secret movement of sub-marines, military officials were loath to reveal clandestine movements of subs, which they would have to do if a transit point officially became part of another country.

In short, it was time, many maritime nations felt, to legally define national reach at sea and draw up rules for navigation, seabed ownership, subsea mining, and commercial fishing. It was also time to establish uniform safeguards against ocean pollution and a way of enforcing them.

The resulting treaty was drawn up in New York in 1973 during the biggest international conference ever held. It guaranteed the oceanfront nations twelve miles of territorial sea and another twelve miles where they could enforce customs and immigration laws. It gave countries a 188-mile-long "exclusive economic zone" at sea where they owned all subsea rights to minerals, but not the surface. Passage there and in international straits would be free to all. The United Nations Convention on the Law of the Sea (UNCLOS), also known as the Law of the Sea Treaty (LOST), was modified over another decade of negotiations and came into force in 1994.

The *Healy* was in the Arctic in 2010 because the convention also established international rules by which coastal states could claim *extra* exclusive economic zones if they could prove that their continental shelf extended beyond 200 miles. And if an undersea mountain range jutted out even farther but started out on their shelf, they could claim more.

Nations had ten years after ratifying the treaty to make claims.

The distribution of these new areas will be "the greatest division of lands on earth possibly ever to occur, if you add up claims around the world," Paul Kelly, an American consultant on energy and ocean policy and spokesman for the American Petroleum Institute, told me.

By summer 2010, Russia had already submitted a claim for an

area the size of France and Spain combined, including the North Pole. That claim had been sent back for more work by the UN committee, which wanted more evidence supporting Russia's assertions. Norway's claim to 91,000 square miles had been confirmed. Denmark and Greenland were preparing claims.

It was, in a way, a race. A very politically sensitive race. At stake was more than any individual country's claims. Nations wanted to have a say in what other countries got too. Building a case pro or con involved obtaining thousands of pieces of information, and on August 25, that involved collecting data about the seamount below the *Healy*.

"When the seamount was created, it brought up sediments. It was created later than the ocean floor. We're pretty sure of that. We thought if we could get a sample and date for that we could use it throughout the rest of the interpretation. It could help us date everything," said Jonathan Childs, who headed the US Interagency Task Force Seismic Data Operations Team. "Knowing the age of formations is critical to establishing geologic framework."

In other words, geologic framework equals legal framework. Legal framework equals recognized claims.

During most of the cruise in 2010 the *Healy* was teamed up with a Canadian Coast Guard icebreaker, the *Louis S. St. Laurent*. The *Louie* was there to help Canada claim territory too. In one area off the border between Canada and the North Slope, both countries claimed the same area but the dispute was peaceful and the hope was that joint work would enable both nations to optimize claims in other areas.

The two ships carried different equipment to do this.

On the *Healy*, 24 hours a day a "multibeam echosounder" embedded in the hull shot 60,000 sound pulses an hour to the bottom. Based on the speed with which the pings reflected back, a computer rolled out data as 3-D pictures of a 110-degree-wide swath of seafloor. Scientists watched on screens in the computer lab, where more monitors and closed-circuit TV provided data about temperature, ship position, wind speed and ocean salinity.

Walk aft to the warren of labs at 4 a.m.; climb down below the helicopter hangar with its basketball court; pass the wet lab, biochemical analysis lab, main lab and enter the computer lab and you'd find a Woods Hole scientist teamed with a high school science teacher on monitor-watching shift, eyeing information crawling across eleven screens—laptops, PC monitors or closed-circuit TV.

The mountain below on August 25 had first appeared on echo-sounder monitors in 2009 as a colossal protuberance in stratified 3-D colors; seafloor in Montana sky blue, base in cobalt, upper regions banded daisy or lava red.

The *Healy*'s other principal piece of research equipment, a sub-bottom "chirp" profiler, dispatched high-frequency sounds to the bottom. These penetrated the seafloor to be bounced back and provide a high-resolution picture of shallow sediment layers.

If you didn't know the chirping was coming from a machine, you might think, walking into the lab, that someone's hungry pet bird nearby never shut up.

On August 25 the *Healy* worked alone, but once the *Louie* fixed a busted propeller shaft bearing she would rejoin us and again begin working her "multichannel seismic system." It profiled deeper sediment thickness.

The *Louie*'s system used three airguns that shot air bubbles into the water every twenty seconds. Collapsing, the bubbles produced sound pulses that penetrated the seafloor. Some were instantly reflected back and others went deeper. Returning echoes were captured by a series of receivers—hydrophones—towed behind the *Louie*.

In clear seas the two vessels could work independently. In light or moderate ice the *Healy* cleared a path for the *Louie*. This was because the *Louie*'s equipment was more sensitive. Breaking ice sounds or zigzagging avoidance maneuvers could interrupt data flow, and anyway, the *Louie* had limited maneuverability when it towed equipment.

Both ships—like the Shell Oil seismic ships in the Arctic—were required under law to carry aboard marine-mammal observers who could alert officers to shut down seismic work or divert route if whales, seals or walrus were in a ship's path.

But no creatures appeared during the job today and finally the core emerged dripping from the ocean. I joined scientists clustered around the pipe.

Was a piece of elephant-country mountain inside?

"There's a protocol of who gets to work with this," Brian Edwards said.

He carried the plastic core liner into the ship's wet lab and labeled it. It contained smeary red-orange slick mud that turned to, several feet up, more granular sediment.

"There's a pecking order of who gets samples."

Brian loaded the tube into a wooden box and placed the box in the *Healy*'s walk-in freezer. It would be shipped to USGS labs in Menlo Park to be "photographed and documented, numbered and tagged."

A "core curator" would control all samples. *Healy* scientists would get first crack.

"You have to apply if you want to see it."

Mud would be measured out, request by request, to scientists. No information would be released to the public about the sample until at least autumn, Brian said. In fact, when I asked White House climate official Heather Zichal about it eight months later, I was told, "I'm not sure what I'm allowed to say. I'll get back to you." But I never got an answer.

The *Healy* headed off to rejoin the *Louie* and conduct more seismic and sonar work.

Jonathan Childs seemed to be speaking about all aspects of the trip—the dating process, the research process, the race for the region—when he said, "It's all about time."

Time. On board, time stretched and weeks or months felt like one continuous day. Not a long one, just one that kept going. Maybe the light caused the feeling. The sun stayed up for 22 hours at a stretch in August, then relaxed into a dim haze. You closed a porthole to shut light out of your cabin at 3 a.m. if you wanted to sleep. Or maybe it was the ship's rhythm, people going to work in shifts day or night. Perhaps it was how the sun *moved*, not up and down but in a circle, never getting too high.

Geologists even made jokes about time since they measured it in eons, not years. "I'll see you Tuesday," was a punch line Andy Stevenson told me. I didn't understand it but he found it hilarious. "Get it? Tuesday? It's a geology joke. It's such a *short amount of time!*"

Mirages appeared and vanished. From the bridge I saw a city on the horizon, made of ice. Distant towers formed a white skyline, except when we reached it the high-rises shrank into a three-foot-high field of ice. But then, ahead! Another ice city!

Sun dogs, small phantom suns, hung in the sky near the main one. Sailors said that sometimes a ship might appear upside down in the distance, floating in the sky. The image would be the reflection of a real ship beyond the horizon. The ships would come closer; one on the sea, the other upside down in the sky. The upside-down ship would sink lower and the ships would touch and then the upside-down ship might abruptly disappear.

Sometimes on the *Healy* the intercom came alive if a polar bear was sighted. Crew and scientists rushed to the bridge, cameras in hand. One bear acted shy, head hung, haunches on the ice as it watched us sideways. Another fled and scientists speculated that it had been darted in the past and feared contact. A mom and cub were spotted trotting north over ice, with a big male in pursuit half a mile back. Males will kill cubs. Another time a lone cub's prints crossed the bow of the ship. The mother must have died.

Diamond dust, bits of floating ice, hung in the air.

Many scientists on the *Healy* were outdoor lovers back home—kayakers, divers, cyclists, motorcycle riders.

The ice sounds never stopped. The cabin I shared with master chief Marcus Lippmann was situated on an upper deck, yet at 3 a.m. the roar of ice pounding the hull made me feel as if I were inside a snowplow scraping past a New York apartment. I heard Volkswagens crumpling against the *Healy*. I heard a sound like the distant thumping of an opening artillery bombardment in the film *For Whom the Bell Tolls*. I heard a slushy whisper like flavored ice oozing from a Slurpee machine. Quick, hard bursts announced the ship's striking "growlers," or small ice. A dozen iron wrenches impacted against steel. A gigantic chain rattled the hull for a full minute.

One night there was a medical emergency aboard the *Louie*. A cook was suffering hideous stomach pains, possibly appendicitis. He needed to be rushed south so a land-based helicopter could medevac him to a Canadian hospital. The *Healy* had no copter on this trip, and the *Louie*'s lacked range to make land. For two days both ships rushed south and the jolts and bumps felt like you were riding in a 30-year-old Chevy pickup with no shock absorbers, barreling up a Rocky Mountain washboard road.

Down in the gym/laundry area, where the assaulting ice lay only a few feet behind bulkheads, riders on the exercycle pitched wildly like John Travolta in *Urban Cowboy* as he tried to stay on the famed Gilley's Club mechanical bronco.

But usually the ride was calm, food plentiful and varied—even if Erin the Canadian ice observer couldn't stand eating it. *They* had fresh-baked bread on *their* ship. *Healy* meals included alleged meat resembling steaks, tasty Mexican fare, fresh or canned fruit, salads, lots of cake. Dry and hot cereals and coffees were always available. So were movies and Armed Forces TV, which broadcast an assortment of network shows, sporting events and regular warnings about alcohol abuse.

The *Healy* was stocked with enough food in the Arctic so that a skeleton crew could survive an entire winter if the ship got locked in the ice. This has never happened, but if it does the crew is supposed to build an ice runway. Then rescue planes are supposed to evacuate most personnel while the rest stay aboard until the ice releases the *Healy*.

At one point during the trip a voluntary hazing was held for first-time crossers of the Arctic Circle. Those who participated did push-ups and jumping jacks on the aft deck, competed in a talent show in the hangar before an audience that threw wadded-up sock balls at bad performers dressed in tinfoil costumes, and during a final ceremony licked chocolate off the bare feet of "King Neptune's Bride"—a female crew member—and sucked a maraschino cherry from the belly button of an overweight master chief clad in a toga. This was King Neptune.

"You think this is hard on you? I'm humiliated," King Neptune told me.

Alleged fun aside, hovering always above the scientific mission were hard political considerations. In the mess on the first day aboard, Andy Stevenson told me, "I'm here to work in preparation for the treaty. You know, the land grab?" Instantly Brian Edwards advised Andy to "Watch how you say things." And explained, "I'm the one who has to deal with the State Department. I know the language." Later Brian demurred at a question on Russian territorial claims. The subject was "politically sensitive" and in his good-natured way, he refused to comment.

At another breakfast, over cereal and eggs, US Geological Survey blogger Helen Gibbons mentioned that the State Department vetted all her entries before they went online. Actually, a whole committee from different agencies did.

Ever present during casual conversations was another subject relevant to any eventual US decision on offshore oil extraction. Many people aboard reported feeling the same tensions that their relatives did back home when the subject of the US economic downturn came up, and the need for the country to find a way out of the current recession. Finding a way out of the recession was a big reason that pro-drilling forces in Congress were pushing for more Arctic oil.

Over lunch Lt. Cdr. Laura King said that the value of her Virginia home had plunged, along with the price of other houses in her neighborhood. Crew members reported less opportunity for advancement in the Coast Guard because, they said, the recession was keeping more people from leaving the service. They feared not finding civilian jobs. Retired Andy Stevenson mentioned that less funding at USGS had resulted in staff cuts and in his being hired as a private consultant for this cruise.

"There weren't any experts on this subject anymore."

If stresses on individuals mirror ones on the nation, then talk revealed a widespread economic angst, the exact factor pressuring Secretary Salazar to open the region for offshore drilling.

But usually the focus was work, and most evenings scientists or crew members gave talks in the science lounge so people aboard could learn about one another's jobs. The jobs often related to a rapidly melting Arctic.

Pablo Clemente-Colón, 55, confirmed to a packed room that "we're having a major meltdown of multiyear ice pack."

"It's not rocket science," he said. "Warmer temperatures result in less ice...Prior to 1989 over 80 percent of Arctic Ocean ice was at least ten years old. Now it is half that."

On another night, the usually quiet Jerry Hyman of the National Geospatial-Intelligence Agency—a former nautical cartographer for the Defense Mapping Agency—said he was on board to observe navigation in Arctic waters but that the agency had very little information about the region. It simply did not, he said, exist.

"But with ice changing, it's possible within the next ten to twenty years there could be lots of traffic through the Arctic."

Jerry's agency needed to prepare.

Melting ice was even making some research dangerous.

"Ten years ago you could go out on the ice and build an ice camp. The ice was so stable you could use that camp for years," Jonathan Childs told the audience one night. "Now conditions have deteriorated. Ice beneath the Russian camp broke up this year. They needed an emergency evacuation. A whole miniature city went down; the Russians got the people out but the equipment cracked right through and it was gone."

The warning had not been lost on North American scientists, he added, showing a slide of a yellow torpedolike automatic submersible drone. It carried sonar equipment for bottom mapping. In 2010, Childs said, Canadian researchers had set up a hut on ice, cut a slit in it and lowered the drone down from inside to conduct work.

But in 2011, he said, "The plan will be to deploy these from the *Louie*. Building another hut on the ice is too dangerous. There's no future in it anymore."

Finally Childs and Brian Edwards gave a joint talk on the Law of the Sea Treaty in which they mentioned something that shocked many listeners, considering the fact that the ship's mission was to collect data for a US claim.

The US Senate has never ratified the treaty, even though President Clinton signed it in 1994.

The US is therefore not part of the process by which land might be awarded under the treaty. Not a single US representative sat on the UN committee analyzing claims.

"If Republicans retake the Senate, ratification will be off the table for the foreseeable future," Childs said.

Make a list of powerful American groups supporting the Law of the Sea Treaty and you'd be surprised at the mix. Presidents George W. Bush and Barack Obama—who do not usually see eye to eye—would be on the list. So would the Sierra Club and the American Petroleum Institute, the Pentagon and the Ocean Conservancy, the Chamber of Shipping of America and the Natural Resources Defense Council.

These groups are not generally allies.

You would think that with such disparate interests backing ratification, it would be easy to accomplish. But year after year a handful of Republican senators have blocked LOST, White House inattentiveness has sidetracked it, and Senate procedural rules have kept it off the table.

In 1995, for instance, after Bill Clinton sent the treaty to the Senate for ratification, Republican senator Jesse Helms, chairman of the Senate Foreign Relations Committee, refused to hold hearings on it.

In 2003 when Republican senator Richard Lugar became head of the committee, he put the treaty on the agenda. The Senate Select Committee on Intelligence determined unanimously that ratification would not harm US intelligence interests, recommended approval and sent it to the full Senate.

Republican majority leader William Frist would not bring it up for a vote.

Treaties have to be dealt with on an annual basis in the Senate. You start from scratch each year. So in 2007 the Armed Services Committee and Select Committee on Intelligence *again* voted to support the treaty. But with the presidential election and economic crisis taking up national attention, according to a report by the Council on Foreign Relations, LOST dropped off the table. LOST was not just an acronym for the treaty. It seemed to be a description of its fate in the United States.

The opposition remained solid throughout this period. Rightwing senators like Republicans David Vitter of Louisiana or James Inhofe of Oklahoma regarded ratification as signing away sovereignty. They were not interested in giving an international body any say in US activities.

"LOST places the interpretation and application of the treaty's terms, such as military activities, in the hands of international courts and tribunals," Vitter said. "This 'lawfare' holds grave repercussions for our rights as a country and provides the United Nations with simply too much authority over American interests and concerns."

Many other Republicans disagreed. Lisa Murkowski, Republican senator from Alaska, fretted during a Council on Foreign Relations round table in New York in 2009 that Senate rules allowed any single senator—just one—to halt progress on a bill under a procedural rule called a "hold."

"In the old days they didn't even have to identify themselves," she told me. "At least they do now."

White House lack of attention has hurt the treaty, she said. "The White House looks for the Senate to lead and the Senate waits for stronger support from the administration."

This was the same complaint I'd heard during the Bush administration when I spoke to staffers at the Senate Foreign Relations Committee, then headed by Democrat Joe Biden. And later that same day—in a town house across from the White House—I'd asked Bush's top environmental adviser, James Connaughton, if he thought the White House was pushing the treaty. He said yes. Would it be ratified that year? He predicted confidently that it would.

It wasn't.

By early 2010, I'd asked Mead Treadwell, head of the US Arctic Research Commission, if he thought the treaty would pass *that* year. He said that the White House was putting its major effort into an arms treaty with Russia. LOST would be ignored or sacrificed politically to get Senate support for the arms treaty.

LOST wasn't ratified yet.

By August, with the United States the only Arctic nation not to have ratified the treaty—although research on the *Healy* proceeded *as if* the US had ratified it—on the political end the nation was so far out of the game to claim territory that the US wasn't "on the field, in the stands, stadium or even in the parking lot," one Coast Guard admiral told me. That much seemed underlined when I'd visited the UN and met Alexandre Albuquerque, a tall, professorial retired Brazilian Navy captain in charge of the UN committee's analysis.

Albuquerque explained that it was possible that the claims of other nations could be ratified before the United States even joined the process. He added that the committee analyzing claims is made up of scientists from claimant nations, many of whom spend half the year *making* claims for their countries, the other half *analyzing* other claims.

When I asked the courtly Brazilian if he saw any conflict of interest in these dual roles, he countered that no scientists analyzed claims from their own countries, so this made the process fairer. When I asked if perhaps there could be some political horse trading going on among scientists—as in, I'll support your claim if you support mine—he conceded that this was possible.

Alexandre Albuquerque said there was a race going on in the Arctic to claim territory.

"It's all about oil and less ice," he said.

———

On September 6 I woke in my cabin, climbed two flights of steel stairs to the bridge, and stepped outside to see the lights of Barrow twinkling a few miles off.

On the science end, the *Healy*'s cruise had been a success, Jonathan Childs said. Research had confirmed that sediments lying outside current US controlled seabed in the far reaches of the Arctic basin *did* originate on the North Slope of Alaska.

Evidence supporting an eventual US claim was growing, but whether anyone at the United Nations would actually get to see it was another matter.

The mapping team would leave and be replaced by incoming scientists who would include a contingent from the Woods Hole Oceanographic Institution. It was headed by Dr. Bob Pickart, whose interests related to climate. Pickart worked with mooring devices that measure ocean temperature, current and salinity. His passion was the study of how ocean and ice interact, particularly during storms.

Departing scientists carried bags to the copter hangar. Barrow lacks a harbor, so we'd have to fly to shore. Stepping outside in float suits for the ride, we were surprised to see a large white cruise ship anchored nearby. Tinged reddish from the rising sun, it looked magnificent and could have easily graced a travel magazine cover.

"Where the hell did that come from?" one sailor asked.

The Coast Guard had not known it was coming and neither had anyone in Barrow. The *Hanseatic*, out of Germany, had crossed the Northwest Passage and stopped in Barrow unannounced. The ship landed tourists, and city residents were surprised to see them walking around, taking pictures.

"Did you see the little green men? That's what I call them. They come out of nowhere," Mayor Itta said at his office the next day.

Two days later—after the *Healy* left—I noticed that a small white sailboat had also arrived offshore, and then an orange catamaran showed up too. One boat turned out to be crewed by Norwegians going around the Arctic Circle. The other by Russians exploring the same route.

It seemed fitting that of all three civilian boats off Barrow that month, none were American, none were expected and all of them put ashore passengers who could have easily walked to the airport, avoiding US customs—which has no presence in Barrow—and flown off to anywhere in the United States.

Mayor Itta said, "The country is not ready for what is happening here."

I asked Itta if Ralph Kaleak's report from the *Healy*—news of what the scientists are learning, and of ice conditions Ralph saw—might help the mayor make decisions about Shell. "Yes," he said. "We're a frontier. There's so much unknown here. Everything I learn helps me see if we're going the right way or not. Without anything to compare this situation to, you can understand my questioning myself. What am I basing my decisions on? Fear of the unknown? Is what I am doing reasonable? I try to put emotions out of it but that is so difficult. We're talking about the Iñupiat way of life."

CHAPTER 6

The Arctic Opens for Business, September

The capital of America's Arctic—the possible jump-off point for its northern future—occupies only 21 square miles, three of them taken up by lagoons. The city was named for Sir John Barrow, second Lord of the British Admiralty, who in the 1800s launched the Empire's failed effort to find the Northwest Passage. The HMS *Plover* wintered over in Barrow in 1853 during that search, at a time when local inhabitants were all Iñupiat. By 2010, the makeup had diversified; 58 percent Eskimo, 22 percent white, 9 percent Asian, 3 percent Hispanic and roughly 1 percent each Black and Pacific Islander. It's a surprise to see Pakistani and Thai cab drivers in the Arctic, but in Barrow that's what you get.

Barrow in the popular imagination tends to be misunderstood or ignored when attention is paid to it at all. *30 Days of Night*, the 2007 commercially successful film set in Barrow, pictured it as overrun with vampires during its dark winter season. Packs of bloodthirsty residents diminished the human population until only two were left.

Journalists—allegedly more factual—have portrayed the city

as a last place on earth. An *Outside* magazine article titled "If I Can Take It There, I Can Take It Anywhere," by Jack Hitt, contained lines like, "In my mind, Barrow was a metaphorical locale, an icy dead end wracked by violence and madness." And "I think of a famous local Eskimo I read about in a town history—Eben Hopson—one of the founding fathers of the modern Arctic. Hopson had a dream...He envisioned toilets across the Arctic."

The assumption that a great gulf existed between the North Slope and the rest of America reminded me of Edward Itta's story about going to a party in Washington on the night of Barack Obama's inauguration. Walking on the street with Elsie, he encountered many people coming up to him with cameras, asking, agape, "Are you *Eskimos? Wow!*"

But what happens in Barrow doesn't stay in Barrow. Drive a ship into the Beaufort Sea and the ripples reach world capitals. It's just harder to see at first.

One reason for this is that in a culture where overt displays of pride, wealth or achievement are discouraged, the guy in ratty jeans driving an old pickup turns out to be a multimillionaire whose decisions create jobs in Texas, Michigan and Massachusetts. The nondescript woman in a dirty parka buying Cheerios in the Value Center is a world-renowned biologist whose work is pored over at the State Department. The unassuming elder sitting at the airport is on his way to Iceland to represent the United States at an International Whaling Commission meeting. What happens in Barrow ricochets across the world.

For instance, two days after getting off the *Healy*, I was sitting with Richard Glenn in his third-story office down the street from Borough Hall.

Richard talking about his full-time job when he wasn't hunting whales, as a board member and vice president of lands for the Arctic Slope Regional Corporation. The Iñupiat-owned, Barrow-based company does over a billion dollars of business annually and is the largest corporation in Alaska, with more than 6,000 employees.

Richard saying casually, "We have petroleum refineries and marketing in the state of Alaska. We have oil-field services in the Gulf of Mexico, Canada and Alaska. We have civil construction. We operate the strategic petroleum reserve for the US government as a contractor. We own land, some available for resource development. Some of our subsidiaries helped put the space shuttle up at the Kennedy Center."

He was understating things. A check of job openings on ASRC's website showed subsidiaries in New Orleans; Denver; Miami; Boston; Seattle; Dallas; Philadelphia; Baltimore; Newark; Ann Arbor, Michigan; Billings, Montana; Atlantic City; the Kwajalein Atoll in the Marshall Islands; Phoenix; Fort Greely, Arkansas, and Omaha, to name a few.

ASRC chairmen have held political fund-raisers in Anchorage and when they do, clients and suppliers quickly write big checks. In state or federal elections, ASRC wields heavy political clout.

"We've recently earned more than $100 million a year in profits," said Richard, who wore new blue jeans and a button-up blue shirt, sleeves rolled to his forearms. He was a compact, athletically built man, bald, with thick glasses, small eyes and a warm smile. "A typical shareholder in recent years got about $5,000 to $6,000 in annual dividends."

From the window we could see the Chukchi Sea, ice-free near shore in September. Every Iñupiat below—moms and their babies, a kid on a bicycle, an old woman driving a four-wheeler to the Value Center, an elder walking to the courthouse or quick mart next door, or to the police station across the street or any of the small houses nearby—was a shareholder. So was nearly anyone on the North Slope with 25 percent Iñupiat blood and, as of 1971, their descendants, no matter where they lived. Richard's family of five meant five shareholders, none of whom had been required to buy stock to get dividends. That's because ASRC isn't the kind of company you can invest in on the New York Stock Exchange. It is impossible for non-Iñupiats to own any stock at all. All members of the board of directors are Iñupiat, selected from different North Slope villages. Many corporate officers are Iñupiat too. Richard often represented ASRC publicly and that year—in Washington or in town meetings held by Secretary Salazar—he was a spokesman urging the Department of the Interior to let Shell drill off Alaska's continental shelf.

"Let me give you some history."

The beginning went back to the 1960s, the days of Eben Hopson, first borough mayor; the man that *Outside* magazine writer Jack Hitt had dismissed as "dreaming of toilets across the Arctic," in the days when there were not too many of them; although if you think about it, imagine removing all of them from *your* city if you consider them an unworthy goal. Hopson's portrait hangs in Edward Itta's office and another one is the first thing you see when exiting the elevator on Richard's floor. A third portrait hangs in the Iñupiat Heritage Center: Eben in a fur parka with a whale swimming nearby, birds flying, ice floating. Eben as an Iñupiat George Washington, wearing thick black glasses as he gazes into the future at his larger dream, Iñupiat political *power*. If the first leg of that journey was the creation of the North Slope Borough to tax oil companies, the second was the establishment of native corporations like the ASRC, which owed its origin to the discovery of oil too.

"We were part of a land settlement," Richard said.

By 1970, he explained, after oil had been discovered at Prudhoe Bay, the problem for companies or politicians who wanted to extract it was that US courts had long recognized that Native Americans had rights to lands they lived on.

"We never lost our land in any battle or in any treaty."

So who *owned* the lands of the North Slope? Companies couldn't drill on it until they knew whom to lease it from, a fact realized then by young, politically active Iñupiats like Eben Hopson and Willie Hensley of Kotzebue—a kid out of college—who won a seat in the state legislature and pushed for native rights in Juneau and Washington, DC.

"In Washington, Secretary of the Interior Stewart Udall ruled that nobody would get any more land until they resolved the issue of aboriginal title," Richard said.

"But Congress didn't want to set up reservations. Native reservations were a failed social experiment. People there weren't acculturating, bettering themselves. There were squalid conditions, so Congress tried to get clever in Alaska and said, basically, 'We'll give you some residual title to lands and cash for the remainder. With that, bring yourselves to better times. Better economic conditions. Better acculturation. Better than reservations in the lower 48.' "

The Alaska Claims Native Settlement Act, passed by Congress on December 18, 1971, established twelve large regional and dozens of small village corporations around the state whose shareholders—who would have to sign up to be included—would be natives. The corporations received 44 million acres of land—roughly 10 percent of Alaska—and $962.5 million in cash. They were supposed to earn a profit for shareholders and also safeguard native cultural and social values. The twin mandate posed a challenge because when selecting lands the new corporations were faced with the choice of opting for acreage suitable for commercial development or areas used for fishing and hunting.

The ASRC received five million acres, some granted immediately, others to be chosen or traded for later. *Choosing* which land you got became even trickier because federal or state designated oil or wildlife reserves were already off limits. Richard personally later helped select some land that ASRC owned across the North Slope.

"Who led our companies at first? One guy was trained as a carpenter. Another was a power-plant operator. A communications technician. They were plucked from different trades and made directors, a director from every village. And because we were on the bloom of oil exploration, some of the first companies they formed did road construction, gravel, assisting the oil industry with ice-road construction. We formed partnerships with other contractors and eventually bought them out."

As a minority-owned company from a disadvantaged population, ASRC received preference in obtaining government contracts, which was why many subsidiaries work on US military bases. By 2010, ASRC was trying to diversify and rely less on government work, Richard said, especially since Sen. Claire McCaskill of Missouri was trying to have the special status changed after some Alaska native corporations—not ASRC—had been linked to graft. McCaskill also objected to the no-bid contracts.

Richard added that if Shell ended up drilling offshore in 2011, ASRC would be a contractor on site with crews and boats to help clean up any oil spills if they occurred.

Richard's large partitioned cubicle was done in a soothing lavender color, with a lilac shade carpet, a comfortable cushioned couch and sitting chair, an executive swivel chair in deep maroon. A baleen wall decoration read in script, "I can do all things through Christ which strengthens me." Eskimo masks, model sailing ships and carved walrus tusks comprised shelf decorations. Reports piled on a coffee table and desk included "Seismic Profiles Across West and East Margins of Hanna Trough" and "Basement Rocks Beneath Northern Alaska and the Chukchi Sea." Another report detailed a business deal with government contracting experts offering to "provide start up industry expertise" if ASRC would give working capital.

Richard himself was a perfect example of the way that North

Slope communities have for years taken in outsiders who want to join them. Many leading clans—the Browers, the Leavitts, the Hopsons, the Ittas—are descended from Yankee whaler/Eskimo unions.

"Whaling crews always need people to help."

Richard had been born to an Iñupiat mother and white father, a technician who worked on the DEW Line, the chain of radar warning systems set up between Alaska and Greenland to alert Washington to a Soviet attack during the Cold War. White civilian employees were not allowed to marry Eskimos then, so the family moved to California, where Richard grew up. His dad returned seasonally to the DEW Line while Richard and his brother, sister and mom stayed in California.

"It was a melting pot there, an accepting society. My mother was a one-person ambassador who would walk into my school and give talks on what it means to be Eskimo." Richard fell in love with Barrow when he visited in summers and spent time with his relatives.

"I discovered a place with no fences. My mom's brothers and my cousins were cool. The horizon was open. In the lower 48 you had gas stations, shopping malls, more gas stations, more shopping malls. Here you could throw a rock almost to the end of town. I jumped in with both feet."

Richard took jobs after school to finance summers in Barrow. By college his roommates in Nebraska told him, "You're wishing your life away," as he pined for the North Slope. By 1986, while working on his master's degree at the University of Alaska, he was walking the Brooks Range for ASRC, looking for mineralization belts so the corporation could obtain desirable properties. Or he was walking Barrow tundra "to help us find our own reliable energy source. Natural gas comes from around 2,000 feet below the surface here, from rock that is cretaceous in age. It's a blanket. We drill, walk a mile away, and see almost the exact same property. We get to know rock layers like people. A formation has traits."

Richard was glib and read people quickly, and his cousins joked about his easygoing, open nature, calling him "the sieve." But he could be political too. He smiled just as warmly when avoiding questions he preferred not to answer. ASRC officials tended to avoid discussing salaries with outsiders, but top executives earn annual sums in the mid six-figure range. They do not, as in many private corporations, get stock options to supplement that.

That September, some people in town were already asking Richard if he would run for mayor when Edward Itta's term expired. But Richard was happy with life as it was. He had no desire to be mayor, he said.

A few miles from Richard Glenn's office, work proceeded that month at a science outpost vital to energy, security and environmental policies in America's high north. It went on at a refurbished old World War II–era naval base closed by the military years before. What I *saw* as I turned off the coast road at a sign reading ILISAGVIK COLLEGE, was a silent, dilapidated industrial-like complex with rows of rusted Quonset huts, corrugated-iron garages, and no people in view. It looked like a location for an Arctic sequel to *RoboCop*, one of those out-of-the-way locales where bad guys hole up and the climax shootout takes place.

Beyond a pair of arched bowhead jawbones—a natural statue greeting visitors—sat a long one-story blue building that stretched out in several long corridors.

The site did not strike a newcomer as a nexus of globally important research but it turned out that the building housed—along with college offices and a dorm—a major reason why *Smithsonian* magazine dubbed Barrow in 2010 "Ground Zero for Climate Change Science."

I walked up steel stairs into the offices of the Barrow Arctic Science Consortium (BASC), founded by Richard Glenn and three friends. It was a nonprofit organization set up to encourage scientific work in Barrow and bring the community and visiting researchers together.

That work done here had national consequences was instantly evident because the hallway was lined with photos of VIP visitors, including Michael Chertoff, George Bush's secretary of homeland security. There were labs and an Arctic science library. Posters tacked up made the corridor a museum of research conducted here by scientists from around the world. Their work underpinned the stories that elders, or scientists on the *Healy*, or State Department sources had been telling me for months.

Spatial and Temporal Variations in the Arctic's Sea Ice Structure was the scientific slant on why many elders had stopped teaching young hunters about ice. Its seasonal properties—researchers proved—were changing.

Measurements of Carbon Dioxide and Methane Fluxes over Arctic Tundra Ecosystem described studies of two key gasses believed by most climate researchers to drive the warming that was melting ice. Measurements taken over polar regions—fed into climate computers—helped researchers predict future climate, and those projections figure prominently in the level of attention with which Congress regards bills to cut emissions in cars, require energy-saving building codes nationally or promote alternative fuels.

Anything related to "bowhead whales near barrow," as another study was called, was relevant to offshore oil extraction. If research showed healthy populations, which they did in 2010; unchanged migration routes; and plentiful food supply, the likelihood would increase that oil companies get to drill. If bowheads declined or sickened, opposition to drill plans would increase and Washington decision makers would grow more reluctant to introduce new potentially dangerous factors into the environment.

BASC-based researchers came from places including the Uni-

versity of California, Harvard, Stanford, the Max Planck Institute
of Germany, China, Russia, the University of Nebraska, and Woods
Hole Oceanographic Institution, to name a few, Director Glenn
Sheehan told me, sitting in his small office. The 59-year-old archae-
ologist had been trained at Bryn Mawr after leaving the Navy,
where he'd run a brig, and was a bearded and deceptively mild-
looking man with a funny, caustic sense of humor and an analyti-
cal eye.

By 2010, over 600 scientists a year visited BASC, Sheehan said.
BASC provided them with labs, rental trucks, airport pickup, snowmobiles, polar bear guards, cold weather gear, all around logistics
help and cell phones that work.

The compound's contributions to Arctic and climate science
went beyond the individual scientists who visit. I got into a pickup
truck driven by Nok Acker, assistant logistics coordinator, and
drove out to the old DEW Line that Richard Glenn's father had
worked on. The radar was nearly automated now, and nearby stood
two small buildings on pilings, connected by wooden walkways.
One was operated by the Department of Energy and the other by
NOAA. Both agencies had chosen Barrow as the perfect location
for installations that collected hundreds of key atmospheric measurements each day informing scientists around the world of Arctic
conditions. The instruments monitored methane, CO_2, sunlight,
wind speed, snow cover, ultraviolet light; all factors that—because
the Arctic helps drive the world's weather—affect crop yields in
Iowa, storm surges in Carolina, temperature spikes from Los Angeles to Kuala Lumpur.

The upshot, when you looked at all the annual visitors to BASC,
was that for thousands of years birds and mammals had migrated
to Barrow and by 2010 Richard Glenn liked to joke that scientists
did too.

Each spring when the eiders appeared in skies over Barrow, the two daily Alaska Airlines flights began disgorging scientists and grad students. During peak summer weeks, they filled a dorm-style hotel at BASC and if space ran out there, BASC offered bunks in a dilapidated house in town nicknamed "The Polar Theater." It was a slum of a place where researchers shivered at night in large rooms.

One September when I was in town, so many scientists flooded Barrow—between BASC researchers, *Healy* researchers going in or out and government or university-based scientists in town to count polar bears—that Sheehan arranged for cots to be laid out in the community center. Researchers slept between the basketball court and climbing wall. The gym looked like a Red Cross shelter but the talk was of *less ice*.

Sheehan told me that prior to BASC's founding, occasional scientists used to arrive in town because Barrow was accessible by commercial air. "People would show up and say, 'Is there a place to stay?'" When the Navy base had functioned, an Arctic research lab was located there, employing hundreds of Iñupiats over the years in support roles, as guides, technicians and outdoor experts. In fact, Edward Itta and friends had grown up earning candy money by selling researchers small tundra animals, 25 cents for a dead lemming, fifty cents for a live one.

When the base closed in 1980, the income was gone. The Navy wanted to bulldoze the place but was persuaded to hand the buildings over to a local native corporation that converted it to science. Part of the base became the college campus and an agreement with the National Science Foundation got the research going at BASC.

"We thought we could provide a place where traditional Iñupiat knowledge and scientific knowledge could meet," Richard Glenn said.

He also hoped BASC would provide jobs to teens and give them a close-up view of science, which might steer them in career choices.

"BASC became jobs for three or four people. Then for a lot of people."

It became even more when whale hunters, working with scientists, began seeing their names as coauthors of scientific papers. "Whale expert to whale expert. Caribou expert to caribou expert," Richard said. "This is a whole town full of ice experts who have a running inventory of conditions in their heads. Put that together with someone else who knows ice and cultural differences disappear. It becomes fun. Like two good mechanics talking about a car."

Whenever I visited BASC, I found that on any given day carrying my breakfast tray filled with eggs, bacon, coffee and fresh fruit to a table in the cafeteria, I'd end up meeting researchers whose work had international ramifications. One morning it was bearded, grizzled Matthew Sturm, world renowned expert on snow who worked out of Fairbanks for the Army Cold Regions Research and Engineering Lab. He crossed Alaska by snowmobile in winters, measuring snow depth. Snow cover reflects back about 80 percent of the sun's warmth that hits it. Tundra reflects back 20 percent. Less snow means more warming, and measuring snow loss is important in predicting climate.

Another time I sat next to tall, blond Swedish grad student Anna Liljedahl, who studied why tundra is buckling, a major reason why, the mayor of Nome had told me, Alaskan roads, bridges and buildings are in danger and why more federal monies may be needed for repair.

Anna said that "in order to decide if there is really any change, you need a long-term record." She was unsure whether current warming trends would continue. Her project had been initiated because of interest in warming but was not designed to confirm or deny it, just to gather information about basic hydrological processes that could affect it.

"That will help us make future projections of climate."

That September two research projects at BASC were particularly relevant to Shell's plans. Both regarded bowheads and were conducted by scientists from the Massachusetts-based Woods Hole Oceanographic Institution, working closely with Iñupiat whale hunters.

Dr. Mark Baumgartner, a tall, youthful biologist, led a team that went out in three motorboats each day driven by whalers, looking for bowheads. When one was spotted Harry Brower Jr. used his harpooner skills to insert "tags"—tracking devices—in their skin.

"We're seeing sea ice shrink as a consequence of global warming," Baumgartner said. "The environment will change, but we don't know how. This baseline study will help us learn how animals forage, how food is organized in the water and why it is here. We also have colleagues flying aerial surveys, mapping distribution of whales."

A sister study conducted by Dr. Carin Ashjian, also from Woods Hole, focused on the bowheads' major food source, a shrimplike animal called krill. Dr. Ashjian wanted to know whether as climate changed, krill would continue to mass in the same places that bowhead had been feeding for years.

After five years of study, Ashjian said, "So far the whales still come every year, whether it is warm or cold. So in that sense if these years reflect climate change, the whales don't seem to care about the degree that's been experienced. If it becomes more extreme maybe they will change their migration patterns. I would hope we'll get long-term data to understand more."

There are places—a Manhattan bar where writers gather, a house on Capitol Hill turned into a club for congressmen, a coffee bar in Milan frequented by fashion designers—that become a nexus for ideas. Scientists met in the tundra cafeteria. They compared notes. The campus on the tundra was such a place for Arctic science.

There was even a new $20-million building to attract more researchers. It sat out on the tundra a seven-minute walk from Sheehan's office, and was again a product of Richard Glenn's work. Sent to Washington by ASRC to try to push for more land-based oil drilling in 2001—he'd taken the opportunity to also lobby for the building. He met with powerful Republican Alaska senator Ted Stevens while Richard's uncle, then the mayor of the North Slope, called Stevens to ask him to go to bat for BASC too.

That year's Senate Energy Bill included funding for the new Arctic research center. It had more lab space, walk-in freezers, new offices, and anyone in Barrow at the library or AC Value Center often—looking at public notice boards—found themselves invited to weekend lectures at the new BASC building conducted by the premier scientists in the world.

I used to laugh when people in New York asked me what there was to do in Barrow, as if nothing interesting ever happened there. It seemed like far more was going on in Barrow than most other places much bigger in size.

Taqulik Hepa was worried about the journalists coming to the North Slope from NBC news. Her office was on the far side of the cafeteria from BASC and she headed the Borough Department of Wildlife Management, which occupied that corridor. Press interest in the Arctic was growing and a New York–based camera crew was flying in to cover a story. Taq's concern was that they would cause the death of thousands of walrus, a prime source of food for several Eskimo villages.

The worry was that the newsmen would accidentally cause many pups to be trampled. And the reason—the same reason the newsmen were coming—was once again, *less ice.*

The late-afternoon September telephone conference was attended by half a dozen staffers grouped around a long table, where a speaker box projected the concerned voice of Nome-based Vera Metcalf, head of Alaska's Eskimo Walrus Commission, which represents natives along the Bering Strait and North Slope, safeguarding community interests.

"If the news crew causes a stampede, aren't they responsible?"

"Can we threaten to arrest the news crew if they come?" asked the voice of Joel Garlich-Miller over the intercom. He was an Alaska-based biologist with the US Fish and Wildlife Service.

Taqulik Hepa is tall, raven haired and authoritative in such a quiet way that I sometimes had to listen hard to hear. And walrus are enormous sea mammals equipped with ivory tusks. A male can weigh 4,000 pounds, a female up to a ton. The tusks are used for fighting or creating holes in the ice for breathing. A major walrus food source is clams that live on the bottom of the shallow continental shelf.

Like bowhead whales, walrus migrate past the North Slope each year, heading north in summer, south in fall. But unlike bowheads, walrus need ice to rest on during the journey.

Males—better able to survive in open water—can do that because they have large air sacs that they can inflate like natural life preservers, but females cannot float as much and calves cannot do it at all.

When there's no ice, calves drown.

So the walrus—in unprecedented numbers—were coming ashore.

"Haulout" is the term used to describe a mass walrus appearance on a beach and villagers at Icy Cape near the village of Point Lay, 140 miles west of Barrow, were reporting as many as 20,000 animals onshore. Normally the walrus would be a hundred miles out.

The eight-by-ten photos I saw—taken from the air—depicted Point Lay as a speck of a village hugging the black-sand beach with a spectacular maze of tundra and rivers at its back, and then, offshore, churned-up waters from thousands of incoming animals, and beaches so covered with the reddish mass at rest that the sand was almost entirely obscured. The animals looked as plentiful as the great herds of buffalo that once roamed America. But they had been driven onto land by desperation and would have to compete for too little food to supply so many mouths.

If they stampeded, thousands of calves could be crushed, and the appearance and drone of a low-flying airplane filled with cameramen could set off the panic.

"They want to overfly," the voice of Joel Garlich-Miller said.

Taqulik Hepa had told me during previous visits that she regularly dealt with consequences of warmer weather to North Slope animals. More polar bears were coming ashore in autumn—lying exhausted on the beach because of *less ice*, for hours—and the borough's vast caribou herds were having trouble reaching forage in winters because snow, which was easy for caribou to graze through, was melting and refreezing as ice, which locked the grasses away. That day Taq was in contact with elders in Point Lay who were going house to house, advising young hunters to steer clear of the beach, to pass up the walrus meat and the ivory recoverable off dead animals, at least until the living animals departed.

Nobody wanted a mass slaughter.

"Those calves are getting hammered," Joel Garlich-Miller said.

But there was not much that the group could do to prevent the journalists from coming. A gap in wildlife laws rendered them powerless. There were penalties for causing a stampede but not for just showing up. A mass death would have to occur before a law was actually violated.

"In the future," Joel said, "we'll recommend that a permit be required for photography in instances like these."

The future. It was uncontested that the future held more haulouts. The recent past had. In 2007, 40,000 walrus had come ashore on the Russia side of the Bering Strait and many calves were killed at Icy Cape during a haulout, John Trent, marine-mammals project leader for the US Fish and Wildlife Service in Anchorage, told me.

"Previously these large numbers of walrus coming ashore in Alaska were unheard of," he said.

It was clear to all present that the 21st century world of the walrus—with less ice and more ships coming, and more oil men, tourists and news people—would be more hazardous to the animals. The group decided to try to divert the journalists by offering aerial footage of the haulout and a press release pleading for cooperation.

It worked. Weeks later I'd learn that the Icy Cape haulout was over and the walruses had resumed their migration.

"But this will be happening more," lamented John Trent.

Newsmen weren't the only media focused on the North Slope that September. If Hollywood can be considered a reflection of mainstream global interest, commercial filmmakers had arrived in the Arctic too. A major movie set in Barrow and planned for 2012 release was scheduled to begin shooting. Casting agents in town looked for extras. One of the film's stars planned to speak out against Shell.

The Universal Pictures $30-million project was *Big Miracle*, and it was based on a real event that had happened in Barrow, a joint US-Soviet rescue of three gray whales trapped in the ice in 1988.

In the film, Drew Barrymore would star as a Greenpeace volunteer who enlists rival superpowers to save the whales. The script reflected a traditional argument between Iñupiats and some environmentalists. It called for Barrymore to tell a local whaling captain that it was wrong to kill a whale and its "babies." The Iñupiat extras would then cheer the captain's reply, that whales feed his family and the whole village's babies.

CAPTAIN: What you are saying is ridiculous. You're a white girl. Go back to California. This is Iñupiat country.

DREW: You don't need to hunt, not when you get stipends from oil companies who have enough money to buy all the food you need.

CAPTAIN: Those stipends last a few months. One day the oil is going to run out. And when that happens, who will feed our children? Will you?

In real life Greenpeace was not particularly beloved in Barrow. The first story I ever heard at the Wildlife Department was about the organization. "They contacted us and said they were doing research on hunting," said Michael Pederson, subsistence research coordinator for the department. "Then the ship showed up and put out a big banner saying 'stop the drilling.'" Pederson wondered if Greenpeace had merely given lip service to research as a trick to win friends. At the time he agreed with the group's overall goal—no offshore drilling—but not the lack of candor or the confrontational tactics. "It's not our way."

In summer 2010 there was also resentment in town about casting for the film. The evening of the walrus haulout meeting, I was invited to meet an English composition class at the community college, and it began with students who had tried out for parts complaining that lines in the script attributed to Eskimos made them look stupid.

"Lines like 'What's an ATM machine?'" one girl said.

In addition, everyone present knew that actor Ted Danson—cast in the film as a win-at-all-costs oil man—was planning to testify against Shell at upcoming public hearings. He was a board member of Oceana, a global conservation organization based in Washington, DC.

Danson's opposition to offshore drilling was widely known in

Barrow. "Oil and water don't mix," he'd told the US House of Representatives Committee on Natural Resources in hearings in 2009. And, "Air pollution from offshore oil rigs poses a health threat." And, "The firing of air guns during oil exploration sends such strong shock waves across the seabed that it is believed capable of causing marine-mammal strandings."

In Anchorage, Shell people were not too pleased with Ted Danson's activism. Curtis Smith—offered a chance to play Danson's PR man in the movie—refused because Danson was speaking up against Shell.

Unlike Shell, Universal Pictures knew its Arctic plan for 2011. It had all the permits it needed. The film would be completed and Barrow would appear on screens around the world in 2012, whether or not Shell got to drill.

Edward Itta had been invited in late September to address the nation, to testify before the Deepwater Horizon Commission in hearings to be televised internationally. The commission had widened the scope of its investigation into the Macondo well disaster to include Arctic offshore drilling and whether that was a good idea.

What Edward said would be heard by hundreds of millions of people, by oil companies wondering whether to invest on the North Slope, by decision makers in the White House and legislators in the state capital. Once again he had the ear of the "big boys," as he called them, but he had no power to make the actual choice.

Itta had talked to Secretary Salazar on September 2, when Salazar flew into Barrow to hold a public meeting in the Borough Assembly Room, downstairs from the mayor's office.

Itta had met Salazar at the airport and, showing him around town, urged him to allow more *onshore* drilling on federal lands, but that plea—that the feds open up ANWR to some drilling—was like beating a dead horse. Itta opposed Shell exploration of the Chukchi leases, he said, but because he also thought offshore drilling inevitable, he said that he hoped industry and government would work together "to allay our fears and concerns."

Itta added a request that he'd not made in speeches. He *wanted* to know whether oil really lay offshore. One of his concerns was that after years of fighting over the leases, exploration wells would turn out dry, and all the precautions he was fighting for would, in the end, do nothing to provide the North Slope with income from new sources of oil.

He told Salazar, "Help us determine once and for all that industry has found something out there that is worth pursuing. I'll support offshore exploration if it is done right."

"I think the secretary was surprised to hear that," he told me.

Afterward, Salazar got an earful in the packed assembly room, a small and comfortable venue with auditorium-like fold-down cushioned chairs and photos of assembly members and Itta on the wall. Itta might be the most powerful voice on the North Slope, but he was far from the only voice.

Sixty percent of Itta's constituents had told pollsters they opposed offshore drilling in 2008. It was an issue that divided families, villages and friends. Itta himself had told me that he'd had many conversations with his family to explain his position.

On that day, Thomas Olemaun, executive director of the Native Village of Barrow Iñupiat Traditional Government, opposed offshore drilling, using the same phrase the mayor did to underline his argument: "The ocean is our garden."

Olemaun told me, "There's no way to clean up an oil spill in the Arctic."

George Edwardson had been a mining and petroleum technician and one of the first Iñupiat pipeline workers when the Alaska pipeline was built. His opposition to offshore drilling had not changed since 2009, when he'd told another hearing, "I understand the industry. The US has to make a baseline [science research] in the Arctic before any development."

The little sigh of relief that Edward Itta had issued when the Arctic offshore drilling suspension had been announced in June had by September turned back to concern. A grinding and almost universal sense of powerlessness marked everyone awaiting a decision from the secretary of the interior. There was no telling *which* meeting, *which* memo, *which* testimony might sway Salazar's call and therefore no interested party could rest. Edward hadn't spent more than a few days at one time at his fishing camp in years. The oil fight had disconnected him from a primary source of his strength, the land. Friends told him that they never saw him out on the tundra anymore. But he was afraid to rest.

Would *all* drilling be stopped? Would Salazar go along with the big environmental groups? Itta had expressed his fears of this to Salazar. He'd told the secretary that the environmentalists "would shut everything down if they had their way. That would be the end of life up here in this day and age."

The frustrating bottom line was, Itta had found the Obama administration more accessible than George W. Bush's. But he still had no idea what Salazar was thinking.

Would the testimony before the Deepwater Horizon Commission be the thing that finally broke the logjam?

As usual, contemplating it he was not sharing worries with his family but with his old pal Bart Ahsogeak.

"Every time he goes to DC he calls me up and says, 'Am I making the wrong move?'" Bart said. "And I tell him, you're representing us. Don't get mad. *Compromise.* Play with them. If they start writing things down, look at it. All the talk—the give and take—means nothing. But if you don't like what they write down, add that one little word."

"And Edward tells me, 'I'm scared if we lose. I don't know how I'd face the people.'"

In Anchorage, Pete Slaiby was getting ready to address the commission also, on the same day as Itta.

Itta looked exhausted. "Sometimes I think, God bless America but let's get into the new age already," he said.

CHAPTER 7

Alaska, Washington and Big Oil, September

Back when Edward Itta was in high school, in the 1960s, oil geologists flying over the northern foothills of Alaska's Brooks Range noticed something that made them sit up straighter. The scene below—the folds in the earth, the layers of exposed rock, the boulders scattered around reminded them of the Zagros Mountains of Iran, major source of that country's huge oil deposits.

The geologists knew by that point that some oil existed under Alaska. Eskimos had burned seeps for fuel for hundreds of years, and the Navy—starting in the 1940s—had contracted with Husky Oil to explore for oil in the National Petroleum Reserve, an immense area south of Barrow, designated federal territory by President Warren Harding between world wars when the country sought a domestic fuel supply for the armed forces. Those wells hit oil—not enough to justify large scale extraction, but enough to suggest that greater quantities might lie elsewhere on the Slope.

Also, by the 1960s oil had been discovered in the Cook Inlet in southern Alaska, so, "Alaska was on the map for oil and gas," Tom Homza, Shell's top geologist in Alaska, told me.

The 1960s exploration focused on the Brooks Range foothills. "North of that was tundra. No mountains. Few outcrops. Entirely coastal plain."

Various companies sank wells, seeking four crucial conditions for a strike. *Source rock* was sedimentary rock capable over millions of years of generating hydrocarbons, squeezing and heating ancient organic material into oil. *Reservoir rock* meant porous rock to store oil inside after it was formed. A *trap* was a large containing structure to keep the oil from draining away, and a *cap* would keep it from percolating up.

"When the foothills were drilled, source and reservoir rock was there, and cap rock, but the traps were leaky," Tom said.

It looked like a bust. But then an Atlantic Richfield crew was required to drill one last well under their contract. They also needed to get their rig home, and the best way to do that was by water, which lay north. Atlantic's engineers figured they'd drill their last well on the way out.

"It was the what-the-heck well," Tom said.

That well struck oil at Prudhoe Bay in 1968. It would turn out to mark the largest field ever discovered in North America.

To envision what the drill bit hit, Homza said that the oil "was not in a great cavernous space like a swimming pool. When you're talking about reservoir rock, it's rock you might pick up along the side of a road, a piece of sandstone. It's porous and can store hydrocarbons. Up to 25 percent of that rock is available to reservoir fluid."

The Alaska oil rush was on, and the state, which owned the land, was the beneficiary. Once the 800-mile-long Trans-Alaska Pipeline opened, income poured into Alaska along with thousands of newcomers seeking employment. At the height of the rush Alaska provided 25 percent of US oil supply. Revenues were so high that Alaskans paid no state income tax and even received annual dividend checks from the state. They still do.

Alaska had become Kuwait of America.

Was it any wonder that by 2010 with the pipeline drying up a slow drumbeat of panic was taking hold throughout the state? The clock was ticking on people's jobs, mortgages, college tuition and health insurance payments. I attended several meetings in which federal officials sought public input over drilling offshore. In Barrow, speakers feared it even if they wanted it. But in Anchorage, one night I heard 30 people testify before a single voice opposed offshore oil at all.

That was no surprise as Anchorage is the Houston of America's north. Oil money sustains the city in the same way that Wall Street profits drive New York, movie money defines Los Angeles, a Ford stock graph mirrors the rise or fall of Detroit.

Crude oil flows from Prudhoe Bay to be refined into Anchorage's roads, homes and commercial buildings. The 22-story Conoco-Phillips Building, tallest tower in Alaska, completed in 1983, cost $65 million in oil profits; the Egan Civic and Convention Center was paid for with $27 million of petrodollars. Hydrocarbons built the $23 million expansion of the Anchorage Museum, the Center for Performing Arts and many buildings at the University of Alaska.

"Oil money," Pete Slaiby grunted one day as we walked in for a hearing at the university, when I complimented the beautiful auditorium.

All the proud monuments of the city.

In September of 2010, as the nation waited to see whether Secretary Salazar would continue the drilling suspension in US waters, any federal official flying into Anchorage to gauge the mood of the people found themselves thrust into the battle over oil from the second they stepped off the plane at Ted Stevens Airport.

Strolling through the terminal, any visitor could not help but notice one gate completely dedicated to ConocoPhillips charter flights. Roustabouts awaiting the next flight to Prudhoe Bay (only men were there every time I looked) did not look like business types. There was no suit in sight, just knapsacks, work boots, thick flannel shirts.

Big Oil got its own private gate in this airport!

And then, reaching the luggage area, it was impossible not to be dazzled by the enormous wall photo of a polar bear mom and cub. The World Wildlife Fund ad asked for help in protecting endangered Arctic animals, the same ones environmental lawyers told me they would use in court to try to stop Shell in 2011 should federal agencies award the company offshore permits that year.

"Securing the future for the Arctic," the ad said. "Be part of our work." An insert showed two native girls, the inference being that WWF speaks for that entire community.

Big Oil or Big Environment? Choose!

The talkative cab driver who took me downtown one day explained, as we drove along Minnesota Avenue, that oil paid for the plows that cleared away snow in winters. That shoppers in passing strip malls included lots of "slope wives" left at home when their husbands flew north to work at Prudhoe Bay. That the street vendors selling reindeer hot dogs during the annual February dogsled races on Fourth Avenue—"Fur Rendezvous"—or the Alaska Trappers Association reps auctioning wolf pelts outdoors, or the tour company operating trains to Denali National Park all did better in flush times.

It does not take long in Anchorage to begin relating even the most casual sights to oil revenues. The young woman spotted coming out of Nordstrom on D Street with her new spring wardrobe and the suburban dad taking his kids to buy ski parkas—stopped on the street—turned out to work for law and accounting firms servicing oil.

Fran Ulmer, chancellor of the University of Alaska–Anchorage and also a member of the Deepwater Horizon Commission, asked by me if she believed it in the national interest for Shell to drill offshore in 2011, demurred due to the stakes.

"Let me tell you why Alaskans can't answer that question. We are deeply conflicted. We are all dependent on gas and oil for our well-being. We shouldn't trust ourselves to be truly objective . . . Our economy would crash and burn if the pipeline stopped, because one-third of our economy is oil and gas, one-third is federal spending and one-third is everything else."

Oil and oil-service companies are everywhere in Anchorage. That September when Pete Slaiby went to work each day at 3601 C Street he was close to Exxon Mobil at 3301 C and BP Exploration at 900 East Benson and Eni Petroleum at 3800 Centerpoint and even North Slope Borough offices a few long blocks away. Within walking distance sat the Petroleum Club, a dim, functional gathering place nestled on a C Street ground floor, in a building sheathed in dark glass blocking outside views of the drinking, sports watching and fund-raisers inside. No one comes to the Petroleum Club for the décor.

"Anchorage's most prestigious private club," as it advertises itself on the web, was described by Mead Treadwell, who had resigned as US Arctic Research Commission chairman by September 2010 and was running for Alaska lieutenant governor, as "a place where the chattering class and oil industry get together. A dinner place that looks like a mausoleum."

"We don't have country clubs. We have the Petroleum Club," he told me.

And at the club, a regular topic of conversation that month, as usual, was another building a ten-minute stroll away, this one occupied by the federal agency responsible for selling Arctic offshore leases.

The offices of the Bureau of Ocean Energy Management, Regulation and Enforcement, or BOEMRE, jokingly pronounced "bummer" by just about anyone not working for it, sat so close to Shell that an enterprising oil executive with binoculars, having situated himself at a tenth-floor window in the Frontier Building, could scrutinize the faces of people walking into the Bummer Building as he tried to identify rival bidders for leases as an auction drew near.

This is what happened in February 2008 before one of the biggest underwater oil lease sales in US history.

Anchorage Shell staff still joke about the Houston exec who stood at a window, hour after hour, trying to spot competitors the day before the blocks went up for sale.

One big reason for the concern was that oil lease auctions are different from, say, a New York art auction, where rivals coveting the same Renoir painting are allowed to bid against each other until the highest offer wins.

In oil lease auctions a bidder gets *only one chance.*

Companies submit their sealed bids—with the name of the block on the outside—before a sale, and in Alaska this happens at the BOEMRE offices.

"You can't offer more once an auction begins," said Shell's Chandler Wilhelm of Houston, who'd helped determine the company's bids in 2008. "When we figure out bids, we always recognize that we could be surprised."

Surprised? The whole explosion of interest in the offshore North Slope had been a surprise. Surprise was the understatement of the new century. After all, Shell had previously *owned* leases in the offshore Arctic in the 1980s and 1990s. They'd drilled wells there, packed up their gear and left, figuring there was gas there, which was not that profitable then, but no appreciable oil.

A 2005 auction, in which Shell returned to the high north after fourteen years away and picked up leases in the Beaufort Sea—the ones it hoped to drill in 2011—had been ill attended. At that time word was not yet out that a bonanza might lie beneath the continental shelf.

"We won many blocks in that sale," Chandler said. "Afterwards

we celebrated at the Marx Brothers Café [a popular Anchorage res-
taurant]. A special menu read 'Welcome Back Shell Oil.' "

But then the US Geological Survey estimated that 25 percent
of the world's oil might lie in the Arctic, plus Shell did new seismic
work, and so company officials in Houston like Dave Lawrence—
who would eventually send Pete Slaiby north—sat down to "take
a new look and gauge our level of interest in the Alaska play, the
reintroduction of the Chukchi Sea," he said.

At first it was just a look. Down in Houston, Dave assembled a
team of geologists and geophysicists and they took out the old well
logs—hard information on what Shell had found last time—and
also new seismic information and new ways of interpreting it, computer modeling techniques that had not existed in the 1990s.

Sitting in a conference room Dave's eyes flicked between seismic
maps on the table, unrolled reams of printouts showing wavy lines
representing sediments, and the analysis of exactly how old those sediments were. They all tried to figure out what kind of organic material
had long ago flooded the basins from prehistoric deltas and whether
temperatures had risen high enough down there, and pressures had
squeezed that material enough, to create oil. And when, eons ago, the
source rocks down there had begun to expel gas or oil. Had those
subsea structures come into existence *before* oil could form, so it
would trap it, or *afterward*, which would mean no oil was there?

"Hmmmm."

Dave Lawrence studied the old logs and saw that in the 1990s,
drillers had found gas but as the drill bit sank lower, *the gas got
oilier*.

Which meant there was a chance that oil sat under the gas.

"If it was oil, it could be big."

Dave felt "just a glimmer" of excitement at first, but it became
a drumbeat as more meetings occurred, in Houston, and in the
Hague. Dave needed to see actual old samples. Shell saves samples

from every old well in gigantic warehouses around the world—core barns, Dave called them—in the Netherlands and Houston, at the University of Texas and in Alaska. Deepwater cores are stored in a frozen state because sediments inside are not consolidated enough to stick together. Shallow-water cores, like the ones from the Chukchi Sea, do not need to be frozen.

By then weeks had gone by since the reevaluation had started, and to Dave, shuttling between Houston and Europe, the excitement grew. "We had cores. We had fluids. They were matching up but it wasn't enough. We looked again at *pressure* in the old wells and matched modeling predictions from the seismic data with what we'd actually found."

Dave's a quiet guy, not a shouter, and when the Eureka moment hit he caught his breath.

Dave muttered, "Oh Lord!"

By the run-up to the 2008 lease sale Dave figured that Shell was sitting on a huge secret. It was time to figure out how much to offer for the leases and that made things trickier because of the single-bid rule. It was possible that similar conclusions about the find had been reached by competitors who were also active in Northern Alaska.

"Especially ConocoPhillips."

How much should the offer be?

That was when Chandler Wilhelm came into it. He was the money man in Houston, the person who would play the bidding game. Offer too little and he wouldn't get a second chance. Bid too high and he'd win the lease but look stupid because his nearest rival offered tens of millions of dollars less.

Or worse, nobody bid on that lot but Shell.

Try explaining *that* to your boss.

"Lease sale preparations are enormous," said Chandler. "People work sixteen-hour days, seven days a week for months."

They're also top secret. "We're told not to talk to strangers about business," Tom Homza said.

"During lease sales companies rent hotel rooms to strategize. They try to bribe janitorial staff. They go through each other's garbage," said Mike Macrander. "We don't throw *anything* away at those times."

Fear of industrial espionage is not just paranoia. "When we first moved into the Frontier Building, five bugs were found in our walls," Susan Childs said.

In the end, bids would represent years of secret analysis, guesswork, gambling, billions in profit or loss.

Rival bids for the same lease might differ by a few dollars, Chandler knew. At other times there might be a vast spread.

And as in an Academy Awards ceremony, all bidders would learn who won a lease at the same instant.

By 2008—year of the sale in the Chukchi Sea—word was out of the possible huge find down there.

"We took incredible security precautions. E-mails pertaining to bid proposals were encrypted. We sent two people with bid envelopes to Anchorage so if one got bumped from a flight or missed it or got into a car wreck we didn't lose the ability to participate in the sale," Chandler said.

"Going to that sale was as exciting as attending the Ali-Frazier fight," Macrander said.

The auction was held at the $39 million Z. J. Loussac Public Library, a modern, brightly lit building dominating a corner lot the size of a small college campus, at 36th and Denali, within view of Shell, built during the oil boom.

On that fiercely competitive day the 232-seat Wilda Marston Theatre—one of two theaters on the ground floor—rapidly filled to capacity with oil execs, reporters and observers while, outside, environmentalists in polar bear costumes protested the sale.

I'm glad I don't have their job, Curtis Smith thought, eyeing the garb as he walked in, grateful he could wear a regular suit.

Inside, John Goll, head of the local office of MMS, which would later be renamed BOEMRE, occupied the stage like a host at the Academy Awards while behind him a big screen identified lots up for auction by number, and an assistant wheeled out baskets containing sealed bids by lot.

"My heart was beating hard," said Curtis.

One by one Goll calmly read off lot numbers. It might have been a farm auction in Iowa but these particular parcels lay hundreds of feet below polar ice. The executives sat scribbling bids down, waiting. Shell and ConocoPhillips were running head to head on several lots.

"I knew my job for the next few years could be on the line," Curtis said. Indeed, winning companies would have to hire more employees in order to proceed with drilling. Losing companies might downsize, shedding staff.

Companies with winning bids would also have to pay 20 percent down *by the close of that business day.* "We have all those wire transfers ready to go," Chandler said. Bidders had been authorized by their companies to wire tens of millions of dollars to the US government.

"We look at our remote. We double-check the numbers. We're on the phone with Houston. I give our finance department instructions and they send the transfers."

And yet after all the hoopla, *after* a bid was accepted the government could *still reject it.* The rules stated that BOEMRE would do its own analysis of fair market value *after* the auction, using the bidding company's data. If a company had tried to lowball the government, BOEMRE could hand the payment back.

"We had two uncontested bids rejected in that sale. The government deemed our bids inadequate," Chandler said.

Oil lease income is the second largest source of government

revenue after income taxes in the United States, so lease sales are serious.

And so on February 6, 2008, there were gasps in the audience when huge bids were announced, high fives exchanged by successful bidders. But onstage there were no showman flourishes. No Hollywood-like announcements starting, "The winner is..."

John Goll simply opened envelopes and calmly announced bids. The winners were obvious.

"You don't even know if any oil will end up being there in the end until you drill," Goll told me later.

If the waiting got too anxious, any audience member could walk upstairs to the library, into the first-floor Alaskan fiction collection and flip through classics like Jack London's *The Call of the Wild*, which begins, "Men, groping in the Arctic darkness had found yellow metal... thousands of men were rushing into the northland."

Hmmm, bad choice for a bidder wanting distraction.

Another choice was Rex Beach's *The Spoilers*, published in 1905: "It's in my veins, this hunger for the North."

No distraction there either.

Shell forked over roughly $400 million in 20-percent-down wire transfers that day and afterward there was a big celebration party at the home of Rick Fox, who headed Shell's Alaska Venture at the time. But by September 2010 the heady days were over. Shell had not officially dropped its plan to drill in the Chukchi Sea yet ("We wanted to keep the pressure on," said Curtis), but Slaiby was pretty sure the Chukchi leases had no chance of going forward in 2011. Between ongoing legal challenges, the Department of the Interior halt in drilling and the still-needed EPA clean-air permit, it looked impossible that the drilling would occur even three years after the Chukchi leases had been purchased.

"For years we've been a bridesmaid, never a bride," Slaiby would say in a frustrated way.

Slaiby's hopes were pinned more on getting permission to drill the five-year-old leases in the Beaufort Sea. Fewer challenges existed there. More research had been done on conditions. And if the Beaufort opened up, maybe the Chukchi—biggest hoped for prize—would be next.

But *still* no word had come from Secretary of the Interior Ken Salazar three months after the drill suspension began. A clock was running because leases are only good for ten years. After that they revert to the government. A company can ask for an extension, but Shell had not done so as it would trigger another bureaucratic permissions process, and that could backfire. Also, as the suspension was ongoing, it was unclear how much more time Shell might need.

Up in the Frontier Building, the mood swung from hopeful to glum, soldierly to depressed.

Curtis said, "It felt impossible that we'd ever get the project off the ground."

Aggravating Shell's thinking was the fact that Department of the Interior officials kept changing the way they referred to the maddening delay. It was a "moratorium," same as the oil-drilling halt in the Gulf of Mexico. No, it wasn't a moratorium, because the word "moratorium" had legal ramifications. It was a stoppage. No, a moratorium. A suspension. A halt.

"Don't call it a moratorium," a Department of the Interior public relations rep snapped at me over the phone.

I picked up the newspaper. Secretary Salazar said the Department of the Interior had issued a moratorium.

One reason for all the secrecy and wordplay, one highly placed federal official told me on background, was that "any decision the Department of the Interior takes is probably going to be litigated. So they speak carefully. If they don't build a rational explanation for decisions, the people in black robes will make the final ones.

"If you don't follow the recipes and create a record, decisions are kicked out of the Executive branch, into the Judicial branch."

In the Frontier Building and in Houston, aggravation was exacerbated by the thought that if Department of the Interior leaders couldn't even decide what word to use in describing their policy, how would they ever figure out what the damn policy was? Shell wasn't the only anxious party in Anchorage. Mead Treadwell, in his bid for lieutenant governor, had released TV commercials showing him walking beside the Alaska pipeline. "It's only running one-third full, and Washington's policies could actually shut it down," he said. Alaska's bounty was drying up and new supply was needed. Mead's poll numbers were rising.

Conversations at Anchorage dinner tables those days waxed nostalgically about the good old days when the pipeline had been full. Curtis Smith and his wife, Jody, had been in high school back then. At their suburban home in the foothills, he often invited coworkers over for post-work drinks, and over a mint-flavored Maker's Mark on ice, the ex–TV newsman would recall with a smile that even the high school kids had so much money back then that each grade spent "tens of thousands of dollars in social funds."

"Heck, if you got a shoeshine you tipped the guy ten bucks."

But Anchorage high school kids weren't splurging now. And in the office Curtis made frantic preparations for an event that Slaiby hoped might budge the elephantine federal government and help end the moratorium…uh…halt…uh…suspension on Arctic drilling. On September 27, Pete was scheduled to testify in Washington before the Deepwater Horizon Commission, on the same day as Mayor Itta.

"Pete told me to start preparing 45 days before the hearing," Curtis said. "He said, 'These next days will be the most important in your career.'"

Curtis had to help ready Slaiby for the big "showstopper"

questions they knew would be coming: Was enough scientific knowledge known about the Arctic to permit exploratory drilling? ("Yes," Pete told me.) Since the Coast Guard had no base in the Arctic, how could it get ships there to clean up if an oil spill occurred? ("We're responsible, not the Coast Guard, and the Coast Guard can have supervisors on board.") And anyway, was it even *possible* to clean up oil in ice? ("Tests in Norway show that it is.")

The message would include that Shell would cooperate with any federal agency to make their plan more robust, and only wanted to sink simpler exploratory wells, not more complicated production wells.

The hearings would be broadcast live around the world.

"We needed to explain to the American public that what had happened in the Gulf in no way correlated to a worst-case scenario in Alaska," Curtis said, referring, as Shell officials had since the Gulf explosion, to the massive pressure under the sea that had worsened the *Deepwater Horizon* disaster, and the fact that the Beaufort well, being shallower, would never be subjected to that kind of pressure.

There was no telling, Slaiby knew, whether once the hearings were over the company would get another good chance to make its case in public. A decision could come at any time and might mean a total shutdown. Slaiby wasn't even sure who would make the decision. President Obama? Ken Salazar? Salazar had told Slaiby that he was responsible but that there would be "other people at the table," Slaiby said.

He wondered, What did that mean?

"How did we know how political this process was or not? That was the biggest question," Curtis said.

"Pete's testimony was key."

Another big "if" was whether Mayor Itta, when his chance came to testify, would attack Shell or support it.

"We hoped he might have at least something neutral to say, if not positive, about our efforts."

This hope existed because Itta and Shell had been hammering out a secret agreement over the summer. Itta had approached Shell to ask if the company would be willing to bankroll joint scientific studies in proposed drilling areas. North Slope and Shell scientists would work on projects together related to drilling—noise research, mammal population research, ice research. The purpose would be to make any eventual decisions more sustainable environmentally.

Slaiby had agreed, and by September lawyers for both sides were determining the rules under which research would be conducted. The agreement was fraught with political risk for both men. Itta risked being seen as a Shell puppet, accepting money for the borough. Slaiby was giving up control of data if in the end it showed that oil exploration might do damage. Shell would be unable to stop information release, unable to control Itta's staffers.

There was no longer such a thing as pure academic science in the Arctic. All studies seemed to impact development. What a polar bear ate—if it had enough food...*any bad effect on marine mammals covered by environmental laws*—might show up in a Ninth Circuit Court brief. That's where Shell's opponents tended to file.

Shell's initial funding of the project would amount to $2 million. Decisions would be made by committees composed of appointees chosen by Itta and also by Shell.

Curtis said, "If the mayor showed comfort with us before the Commission, we felt that would buoy our chances of others taking what we had to say to heart."

Shell staffers also believed that Obama administration appointees would be more sympathetic to a Native American view of the Arctic than to an oil company's.

"Given the history of Mayor Itta and his ability to stop our program, we put significant weight on what he would say," Curtis said.

Slaiby—however—decided not to mention the science agreement in public yet. He feared that if he did, Itta might feel coopted and back away from the deal.

The fight over Shell's leases—once barely known outside Alaska and still overshadowed by the Gulf of Mexico—was about to hit the world stage.

What the hell were they thinking in Washington?

Curtis's office at Shell is filled with small happy touches: a model airplane, a huge box of Peppermint Patties available to anyone. He moves through the corridors with a boyish morale-lifting attitude that cheers people. I asked him before Slaiby left for Washington whether hope existed that Pete would do so well in the hearings that it might turn the tide totally in Shell's favor.

Curtis shook his head sadly. "We stopped dreaming like that years before."

Still, all wondered, how would Slaiby do?

And what would Mayor Itta say?

The Russian in Anchorage was amused by American gridlock in the Arctic because it was so different from the bullish attitude back home. He was 75 years old, bald and slow moving, but his mind was sharp and his eyes showed keen twinkling intelligence. His room in the Hotel Captain Cook occupied a high floor. His open laptop sat on his desk and he jabbed his index finger at the video running there, showing healthy-looking men in gym shorts running on a treadmill. His English was slow and measured. It was a week before Pete Slaiby and Mayor Itta would testify in Washington.

"First we made nuclear bombs," said Evgeny Pavlovich Velikhov.

Velikhov—a lead speaker at the conference downstairs on the opening Arctic—was an ex-winner of the Lenin Prize and was president of Russia's prestigious Kurchatov Institute, its leading research and development institution in the field of nuclear energy.

"Then we made nuclear submarines."

He had also been a Russian arms negotiator and had been designated by the United States as a hero of Chernobyl after he led efforts to clean up that nuclear disaster.

"Arctic is next big place!"

Now, he was promoting a new way to get hydrocarbons out of the Arctic seabed by using nuclear power. In Russia there was no question of *whether* drilling would proceed in the Arctic. The question was how.

From a national security standpoint, the Russian said, when it comes to the problem of terrorists of the future attacking oil infrastructure, or of fighting in the Mideast interrupting international oil supply, "Arctic is more safe than Persian Gulf, yes?

"But in Arctic we have problems, icebergs. If we stop operations in times of icebergs it is not very good, yes? Another problem is oil spill in ice. Recovery? Practically zero," he said. But he did not look worried.

On the contrary, he was smiling. He had an answer! His eyes lit up and his blunt fingers stumbled over the keyboard, stopped, lifted a pen, sketched.

"We need special technology for Arctic, yes?"

Astounded, I realized he was drawing an *underwater nuclear powered tanker*, a kind of huge submarine that, he said, would travel *beneath the ice*, arrive at a sea bottom wellhead, attach itself to piping—as in the sketch—and suck up oil or gas.

The men running on his laptop represented the happy and physically fit staff that would manage the subsea facility.

"Our intention is to put all the exploration and production under the ice. Surface is not friendly."

I blurted, "You're *building* this in Russia?"

"We could in three years."

"Ah, it's just a theory?"

"No! Not theory! The Germans had underwater tankers in World War II. Deliver supplies."

"But those were U-boats. U-boats are small. You can't transport usable supplies of oil in little U-boats."

Velikhov smiled. "Bigger is easier. And now government is starting interest."

Russia, he said, already operated several nuclear-powered icebreakers. Russia knew the value of Arctic oil, he explained. Russian subarctic oil had fueled that country's resurgence as a world power. Russia's new Prirazlomnoye Arctic offshore oil field was scheduled to come online soon, with 35 more wells planned offshore in the Pechora Sea.

Underwater tankers carrying highly flammable liquefied natural gas, Evgeny argued in the hotel room, would be safer than traditional tankers because they couldn't catch fire if there was an explosion. After all, there was no oxygen under the sea to ignite. The subtankers could operate year round without worry about icebergs. They would arrive at an underwater wellhead, couple into the pipes, pump aboard gas or oil, and be on their way to supply the world.

"*Ve-ry* safe."

Coming from someone else these ideas might have seemed like science fiction. But Velikhov was actively involved in discussions with the Russian government about them, and his history as an innovator had gotten him invited to the gathering downstairs—Icetech—sponsored by Anchorage's Institute of the North, started by ex–US secretary of the interior and former Alaska governor Walter Hickel.

Dedicated to the study of Arctic policy, Institute staffers believed the Arctic was opening fast, and it was clear, walking around in the packed ground-floor-ballroom conference location that although the US government was not dealing head-on with polar issues, private companies, usually from other countries, were.

Arktos Craft, based in Canada and Switzerland, was showcasing an amphibious craft specializing in "interface between ice and water." Photos showed hard-hatted crew, in ice-covered water, carrying out an evacuation in the North Caspian Sea. The red crafts could float or use wide treads to rumble like tanks over ice. They had "the ability to maneuver through ice rubble fields and ice/water transition zones while carrying heavy loads," the brochure bragged.

Aker Arctic Technology of Finland announced a joint project with Russian companies to develop a new "oil spill combat ice-breaker" for Sovcomflot, the largest Russian shipping company.

Oceanic Consulting Corporation of Canada promoted "solutions for the Petroleum and Shipping Industries." Glossy brochures showed shots of ships moving through ice fields, and photos of indoor ice tanks for testing hull design. Representatives from US federal agencies, corporations and foreign governments circulated through the exhibition space. The US Coast Guard, Shell, Korea Maritime University, BOEMRE, the Canadian Navy, Rolls-Royce Marine, the Panama Canal Authority and Maritime Helicopters all had people there.

The Cook Hotel seemed an appropriate venue for a meeting designed to discuss an opening region of earth. After all, the building was named for the famed British explorer whose sea journeys in the 1700s mapped opening areas of the globe previously unknown to Europeans—new territory eventually claimed by world powers—back then. Colorful nine-foot-tall wall murals on the ground floor showed scenes including Maoris in New Zealand, which Britain would claim; the South Seas, where they made more claims, and even waters lying within view of Mead Treadwell's glassed-in back porch five minutes away. One mural showed Cook's ship near icebergs in Antarctica.

Conversations in the ballroom went on in Russian, Finnish, Korean.

Speaker after speaker told the rapt audience that the Arctic was in a state of rapid change.

Velikhov, during his turn, informed several hundred listeners that on that very morning, Russia's first portable nuclear power plant was being built to be barged up to the Arctic and supply power to communities there.

"You bring the unit to the customer and make the connection with one or two pipes. Customers not need to have any experience with nuclear power. You drop it off and come back 60 years later and pick it up," he said.

Canadian speaker Arno Keinonen—a naval architect and marine engineer—said, "Bottom line in the Arctic? It is challenging. Potentially deadly. Costly. But it can be done. We are all going to be busy for a long time."

I ran into two US Coast Guard representatives in the hallway outside and asked them what the Coast Guard was doing to prepare for the opening Arctic.

The younger officer looked disgusted.

"Janet Napolitano," he said of his boss, the secretary of homeland security, "doesn't give a shit about the Arctic. We only have one working icebreaker."

Standing beside him, Rubin Sheinberg, a small, bald chief naval architect of the Coast Guard, cautioned the younger man to watch his language.

Then Sheinberg said, "All the other countries are ahead."

Walking into the Deepwater Horizon Commission hearing room in Washington's Reagan Trade Center, Pete Slaiby tried to gauge the crowd. Neutral, he thought. Just riveted people watching to see what he would do.

He was calm, noting the raised dais with commissioners, the table below them where he would testify, the masses of cameras pointed his way. Co-chair William Reilly, looking down at him, was the one who had widened the commission's initial role of investigating the causes of the explosion in the Gulf of Mexico to include the Arctic, a move that had surprised the White House.

"The White House said, 'What are you talking about the Arctic for? That's not deep water,'" Reilly told me. But Reilly had taken the wording of his presidential instructions, to *appraise the future of offshore drilling*, "as an occasion to opine on the basic environmental challenges presented by that question."

From an oil company perspective the Slaiby-Itta session did not start on a promising note. Reilly announced that he was "amazed and disappointed" at the failure of oil cleanup technology to evolve in recent years . . .

Reilly told the world that "oil exploration continues in frontier environments in areas that offer enormous promise for returns as well as risk of catastrophe. How would we respond if a similar disaster [to Macondo] occurred under the sea ice today or tomorrow?"

Fran Ulmer, chancellor of the University of Alaska in Anchorage and a former Alaska lieutenant governor, said she was "very concerned about the implications to Alaska," and asked Thad Allen, former commandant of the Coast Guard, if that service had "the assets that would be necessary to move forward in the future," should an Arctic spill occur. Allen had been head of the US government's response effort in the Gulf of Mexico during the Macondo spill.

"Do we need more than we have now?" Allen responded. "I would say yes."

Secretary of the Interior Ken Salazar issued a statement.

"It was because of what I consider to be the lack of sufficient science in the Arctic and the reality of oil spill response in conditions that are so different in the Arctic than they are in, say, the Gulf of Mexico, that we made the decision to withdraw the leases that had been scheduled to go forward in the Chukchi and Beaufort Sea."

Salazar mentioned the lesser Coast Guard response capability up north. "Reality in the Arctic is also that you're operating in frigid conditions with floating ice."

"Overall, my approach as secretary of the interior has been to go slow, to be thoughtful, to develop additional information," he said.

One wondered, listening, *Then why did the Department of the Interior grant the leases in the first place?*

Next up was Alaska's junior senator Mark Begich, ex-mayor of Anchorage, who weighed in against the go-slow pace.

"I was just listening to Secretary Salazar [say that] in a few months they'll have their report out on the moratorium. Well, a few months—then they're going to take, knowing how the federal government works—another few more months to think about what the report should say or not say and then they'll take another couple of months in reviewing it. Then we're into March of next year, probably at the rate they're going."

Begich warned that continued delay would push oil companies out of Alaska altogether.

"Tell us what the rules are," he said.

Coast Guard spokesman Capt. John Caplis, chief of its Office of Incident Management and Preparedness, spoke of "the tyranny of distance" when it came to getting Coast Guard assets—which were not stationed in the Arctic—to any oil spill that could occur there.

This was irritating to Slaiby and misleading, he felt, to listeners. During the Gulf of Mexico spill the Coast Guard had overseen cleanup efforts, but the actual ships and personnel doing the cleanup had come from industry. It was *Shell's* job to clean up if an incident happened. Slaiby had no problem with that.

In fact, he had for weeks been arranging for oil-spill-response vessels—smaller cleanup boats and a larger ship for storing recovered oil—to be standing by at Arctic drilling locations if Shell got the go-ahead.

"We have our assets under contract."

Slaiby told the commission that a capping and containment system would be built and in place before any drilling would begin, that science research was ongoing in the Arctic Ocean, and that the Coast Guard could easily discharge their role in a worst-case scenario by sending up supervising personnel from their base in Kodiak, three hours away by plane.

"What's the size of the prize...25 billion barrels of oil and 120 TCF of gas...$72 billion in payroll, 35,000 jobs and those jobs are average over a 50-year period, hundreds of millions for Alaska and contractors, extended life of the Trans-Alaska Pipeline," said Slaiby.

When Mayor Itta's turn came, Slaiby's hope rose.

The mayor appeared on screen, in Barrow, and alone at a table, leaning over a speaker phone. He had chosen to testify from the North Slope, where he'd been busy attending to local matters.

"My name is Edward Saggan Itta. I am an Iñupiaq Eskimo, born and raised here in Barrow, Alaska, northernmost community in the United States..."

Well, that much was good, Pete thought.

"In recent years people have become increasingly interested in the Arctic because it is a heat sink for global climate change that is happening..."

Still okay...

"The retreat of sea ice has also opened in vast, new and promising areas to oil exploration...Some groups have been quick to point out certain differences between the Gulf of Mexico and the Arctic Outer Continental Shelf...There's no question that conditions in the Arctic are different in many ways, but those differences are far more daunting."

Uh-oh, Pete thought.

Itta went on to say that the oil spill equipment and technology that Slaiby mentioned had "never been tested in the Arctic in real-life situations."

In this he was voicing a frustration he shared with Pete Slaiby. Witness after witness was telling the commission that more information was needed on the ability of the oil industry to clean up oil in ice. Itta was pointing out that federal rules prohibited a test. He *wanted* a test.

Itta said, "We need to change the permitting system. We need to slow down and measure the impacts as we move out into the ocean. The current plan is a race against the clock when it is not necessary or productive."

Itta concluded by urging that any federal rules for offshore drilling include requirements for pipelines running to shore, so they are buried far below ice.

He never mentioned the science agreement with Shell. Observers were left with the impression that he opposed Shell's plans altogether.

Slaiby, disappointed, said afterward, "I thought he was flat. It's difficult to be engaging over video. Maybe it was that. I had the impression before this appearance that the mayor's issues had moved more toward the agencies rather than towards Shell."

Slaiby headed back to Alaska, telling himself, "Maybe Mayor Itta had health problems that day."

Another surprised listener to Itta had been William Reilly, who had met the mayor earlier that month in Anchorage and come away more than a little impressed.

"I love that guy."

Itta had told Reilly over dinner, "We don't know enough about the Chukchi Sea to go forward." But in the Beaufort Sea, Reilly recalled, Itta said that "Shell has accommodated us by changing their schedule for drilling. The places where they will drill will not interfere with the whales migrating."

Reilly thought Itta supported the Beaufort plan, and when Itta didn't say that in testimony, "I was very surprised. I thought, This is a pretty straight shooter."

Reilly added, "Then I thought, Well, I guess he's a politician. Maybe he'd taken some soundings and this is what he had to do."

On October 6, 2010, in an effort to get the federal government moving, Shell announced that it was scaling back hopes for 2011, withdrawing the Chukchi plan from the table altogether.

Curtis told reporters that although Shell remained excited about the oil below the Chukchi, until ongoing roadblocks were resolved the company would limit its requests to one or perhaps two exploratory wells in the Beaufort.

Slaiby said, "We hope that once we get to a place where people have seen operations we'll be able to slightly scale up."

Shell's revised permit application—changed voluntarily and also along lines mandated by DOI since the Gulf incident—specified that if the company got permission to drill, it would have spill-response vessels standing by. It would bring a containment dome to be lowered over any leak. A second drillship would be present to drill a relief well if a leak or blowout occurred.

Opponents like Betsy Beardsley of the Alaska Wilderness League told reporters, "The government shouldn't bow down to pressure from the industry, but instead stand behind their pledge to let science lead any decision in the Arctic."

The only response from the Department of the Interior was more silence.

Elsewhere in the Arctic things were not so slow.

On October 9—in the Arctic Ocean—Russia's two Arctic expeditions led by the icebreaker *Akademik Fyodorov*, flagship of the Northern Fleet, met to share breakthrough discoveries on their undersea territorial claims. Scientists aboard, like those on the *Healy*, had been working to document them. The Russian press reported that details of the work were "top secret" but that expedition head Andrey Zenkov was thrilled at the results.

"We're not going to ask anybody what we should do," boasted Artur Chilingarov, Russian envoy to the North and South Poles. Chilingarov had once dropped a small titanium Russian flag by minisub at the North Pole.

"The Arctic belongs to Russia," he said.

Russia was hurrying to gather evidence for its submissions by 2013. Canada would also substantiate its claims that year, press reported. Meanwhile the US National Oceanic and Atmospheric Administration announced that the Arctic "continues to warm at an unprecedented rate."

Summer sea ice cover was at its third lowest point since satellite monitoring began in 1979. Sea ice continued to thin, and snow cover duration was at a "record minimum since record keeping began in 1966."

That autumn, days dragged on at Shell in Anchorage. The company was spending money on designing the new oil spill capping and containment system, similar to the one that ended the *Deepwater Horizon* spill, and on sending out engineers to survey routes by which an eventual oil pipeline from offshore might cross onto the North Slope, hopefully to connect to the existing Alaska pipeline.

Shell was paying for copter crews, boats, coordination centers in North Slope villages and Eskimo and commercial marine-mammal observers who would make sure that the Shell boats stayed away from sea animals as they worked.

Soon it would be time to put more contracts in place for the 2011 season. Once again the meter was running.

"Then there's the $3.5 billion we've already put into this since 2008," Slaiby told me in Anchorage. "What's the interest on that alone?"

CHAPTER 8

Russia Moves In, October

While Pete Slaiby and Edward Itta awaited the US government's decision, Russian strategic moves in the Arctic continued at a steady pace.

That's why on October 14 a US Air Force AWACS (airborne warning and control systems) plane rose into the sky over Anchorage and headed north to intercept a pair of incoming unidentified aircraft. Early warning system sensors had warned Lt. Gen. Dana Atkins, commander of Elmendorf Air Force Base, that the speed and course of the strangers meant that they were probably Russian bombers.

It was happening again, Atkins thought.

Atkins's job was to protect US airspace in the far north. As the highest-ranking Air Force general in Alaska, he headed the 11th Air Force at Elmendorf, in Anchorage. As Alaska's NORAD region commander he was charged with detecting, analyzing and alerting US and Canadian governments of air or cruise missile threats.

The incoming planes were 300 miles out and sweeping toward the US Air Defense Identification Zone, international skies close enough to national territory to trigger military interest. They would enter US airspace if they reached twelve miles from shore. Probably, Atkins figured, they would turn away before that if they were Russians. So far they always had. But Atkins had no idea of their intentions—whether they probed for holes in defenses, whether they were on a routine training flight, or whether they had some other mission in mind.

"I don't know what they are doing, when they are doing it or why."

The Russian Tupolev Tu-95 "Bear" bomber is driven by four turboprop engines, giving it long-distance capability. It's usually armed with cruise missiles and tail gunners operating twin cannons. Bears came into regular use in 1952 and have been upgraded since then. Modern versions are expected to serve Russia's Air Force until at least 2040, US military experts believe.

General Atkins told me in 2010, "We're seeing more Russian bombers in the Arctic."

The history of Bear bombers mirrors that of modern Russia's strategic, territorial and military interests around the world. It was a Tu-95 that dropped the USSR's first AN602 Tsar Bomba, the most powerful nuclear weapon ever detonated, in 1961, and ironically in the Arctic, to the horror of enemies of the then Soviet Union.

During the Cold War, Bears flew maritime surveillance, tracking US Navy ships around the globe. Each week, a pair of Bears left Russia's Kola Peninsula on its northwest coast to head for Cuba, their provocative flight path skirting America's East Coast. US fighter jets routinely escorted them as they flew.

In 1988 alone, 57 Soviet military flights were recorded at the edge of US airspace.

Bears only actually entered US airspace once, in the 1960s, and that incident was considered to have been an accident. But the bombers always caused tension when they showed up. When the Cold War ended and the patrols stopped, US military officials breathed a sigh of relief.

Then in summer 2007, Russian president Vladimir Putin ordered Bear flights in the Arctic to start up again.

Some of the bombers by 2010 were showing up near Alaska with Russian fighter escorts. In fact, a North Slope rescue pilot I had flown with said he often heard local pilots speak of Russian sightings.

"This becomes problematic for me," Atkins told the Fairbanks Chamber of Commerce in a speech in 2009.

And so, October 14, the old game afoot, he was sending the AWACS aloft and also a refueling tanker out of Fairbanks. If things heated up, intercepting fighters would follow. Elmendorf based F-22 Raptors could fly at speeds well over 1,000 mph and were the most modern fighters in Alaska. Intercepts sometimes involved sending both kinds of planes as far as 600 miles out. Sometimes fighters went off right away. Sometimes, if Atkins had time to make a decision, he checked with the NORAD commander and even the secretary of defense before sending up fighters. In the Arctic cat-and-mouse air-defense game, it was not always considered politically wise to send up fighters. "There may be some ongoing diplomatic issue elsewhere, so they'll say, The risk/reward is not worth it right now," the general said.

Which of course could be a big reason why the Russian bombers kept coming, to check the US reaction under different circumstances.

Atkins said, "The Russians seem to be showing their presence to bolster their Arctic claims."

Each incident was potentially dangerous. Intercepts often involved midair refueling. Flight paths took US fighter pilots over vast portions of ocean. Accidents were always a possibility. Ditch an F-22 in the Arctic Ocean and even if the pilot survives impact,

Alaska's Eskimos have over 100 words for ice. Most feared in Barrow is an *ivu* (see previous page), formed when large masses of ice collide and suddenly create a mini-mountain range. *(Courtesy of Craig George.)*

EDWARD ITTA: Edward as the mayor of Alaska's North Slope Borough with Coast Guard commandant admiral Robert J. Papp Jr. and Senator Lisa Murkowski; as a boy eager to help whale hunters and at his crew's bowhead whaling camp in the Chukchi Sea. To his right is the small boat the hunters will use. *(As mayor and as a boy, courtesy of Edward Itta; hunting shot, courtesy of Bill Hess.)*

ENTER THE OIL MAN: Peter
Slaiby of Shell arriving at Wiley
Post–Will Rogers Airport in
Barrow, Alaska; explaining
Shell's plans to the citizens of
Barrow and at a congressional
hearing in Washington, DC.
(Arriving and at Congress,
courtesy of Peter Slaiby; at
Barrow, author's collection.)

ABOARD THE *HEALY*: Rear Admiral David Titley, the oceanographer of the US Navy (top), aboard the Coast Guard icebreaker *Healy* in the Arctic. Also aboard the ship, but on a different trip, the North Slope's native observer Ralph Kaleak, wearing short sleeves in 27-degree weather, and Brian Edwards of the US Geological Survey, chief scientist aboard during the summer 2010 undersea mapping mission. *(Titley, courtesy of the US Navy; others, author's collection.)*

THE DRILLSHIPS: The circular-looking *Kulluk*, a floating Arctic-class drill rig, was to stand by during the 2011 season to provide relief in the event of a blowout. To drill an exploratory well in the US Arctic, Shell would use the *Noble Discoverer*, seen here in Dutch Harbor, Alaska. *(Photos courtesy of Shell.)*

NORTHERN LIGHTS: Peter Van Tuyn, an attorney whose clients include the Alaska Wilderness League; Mead Treadwell, former head of the US Arctic Research Commission and Alaska's lieutenant governor; and Harry Brower Jr., the head of the Alaska Eskimo Whaling Commission. *(Van Tuyn, courtesy of Ken Robertson; Treadwell and Brower Jr., author's collection.)*

BARROW LIFE: The northernmost high school football held in the United States, Barrow. Richard Glenn in the ice cellar dug into permafrost behind his home, surrounded by meat that was gathered by hunters and will stay frozen all year round. To honor the mayor (center, below) in his last year in office, Barrow's high school was filled with dancing, speeches and storytelling. *(Football and Glenn, author's collection; Itta, courtesy of Bill Hess.)*

WHERE THE PAST MEETS THE FUTURE: Iñupiats still hunt bowhead whales, which can reach 70 feet in length, in open boats called *umiaqs*. Hunters throw a harpoon on which is mounted a darting gun. The gun fires a small hand-packed superbomb that detonates inside the whale. *(Courtesy of Bill Hess.)*

"If he has to eject and lands in water, that's problematic," Atkins said in his understated way.

"Russia doesn't necessarily adhere to international protocol for flying," he added. "The protocol is, you file a flight plan which is transparent. You talk on radio to air-traffic controllers whether it is in international waters or sovereign airspace. The Russians don't do that."

What the Russians *did* seem to be doing on October 14 was heading south and then—if they did what they had done before—they'd turn west along the curve of Alaska toward the Aleutian Islands in what Elmendorf staffers call an "Aleutian run." The Bears skirt US airspace and return home. But you never know whether what starts out as an innocent Aleutian run will turn out to be different.

"Being just outside Alaska and seeing a Russian aircraft is pretty significant," said Maj. Wade Bridges, a pilot with the 477th fighter group, which Atkins sends on intercepts.

Elmendorf pilots are on call 24 hours a day. Some alerts begin when horns go off and the pilots run for their jets. Others develop more slowly when US sensors pick up the approaching bombers farther away. F-22s may have to fly as much as four hours just to reach the point where the intercept will take place, Atkins said.

Atkins is an intellectually curious man with straw-colored hair who assumed his post in 2008 and whose staffers are under orders to bring him any news they find about Arctic security, climate, ice melt, the Northwest Passage clearing of ice, and oil spills. All could affect his work, he felt. He knew that someday his engineers might be cleaning up oil off the North Slope.

"I have my engineers train by cleaning oil up on a small lake on the base. The ice there is about four feet thick. Small scale," he said, envisioning something larger in the ocean, "but they can at least practice."

Even casual talks in the general's wood-paneled office contain two constant reminders of the way—when it comes to international relations—once-friendly countries can become enemies and switch back again. The reminders are two upholstered chairs that once cushioned President Richard Nixon and Emperor Hirohito of Japan when they met on the base in 1971. Atkins usually offers visitors a choice of chairs. "Would you prefer to sit in the presidential chair or the emperor's chair?" he says.

I sat on Nixon's chair. Atkins, in the other chair, wore a zip-up, olive-colored flight suit.

"The Russians are more active in areas that are disputed," he said that day.

"We've got a big job ahead of us to secure the Arctic in the future."

The October 14 intercept ended up with the Russian Bears flying the route that Atkins had anticipated, skirting the Aleutian Islands and returning home.

Atkins's planes returned safely to their bases also.

In Anchorage and Fairbanks, the pilots waited for the next alert.

But "the military is ill prepared to do anything in the Arctic," General Atkins said. What he meant, he added, was that "we have very limited communication or satellite networks. If I want to talk to my folks, we basically send 'em up with Iridium phones...We absolutely have to improve communications. We probably want to think about fixed bases in the north. Right now I can get planes up there and sensors, but there comes a point where I'll either have to refuel them continually in the air or recover them."

Atkins was frustrated because he felt that most Americans simply have no idea of the vastness and even location of Alaska. He said that his daughter, an elementary school teacher in Spokane, usually asked her students what they knew about Alaska. "Most kids thought it was tiny and located south of Los Angeles. That's how it appears in little boxes on maps."

Atkins added that he told Secretary of State Hillary Clinton of Air Force needs when she stopped by at Elmendorf once and waited for her plane to be refueled on its way to Asia. The secretary of state listened, he said, and suggested to him that "we have five years to do it."

He disagreed. "What it takes in one year to build in the lower 48 takes five years in the Arctic," Atkins said.

"I said to her, if you believe we have five years, we're probably already too late."

Ask a US State Department official why the Russians have resumed aggressive air patrols throughout the Arctic and the answer is that the patrols are "standard procedure" and "stopped after the Cold War because of budget considerations" and restarted because Russian president Vladimir Putin wanted to "show that Russia is back." The patrols are nothing alarming, journalists are assured.

On the record, any belligerent talk about the Arctic coming from Moscow is "intended for a Russian domestic audience" and "not to be taken seriously," was the response I got at Foggy Bottom.

But the 2008 Coast Guard report that said that the Russians have "resumed strategic bomber flights over the Arctic for the first time since the Cold War," took them seriously. So does anyone reading Russian newspapers, which present a sharper view of Russian attitude.

Russia's Lt. Gen. Vladimir Shamanov—for instance—told reporters from *Red Star*, the Russian Army newspaper, in 2008 that "after the heads of several countries disputed Russia's rights to the resource rich Arctic Ocean shelf," the military immediately began adapting training for units that "might be called upon to fight in the Arctic."

A year later Russia's national security strategy—approved by the president—identified the intensifying battle for ownership of oil and gas fields near its borders as a source of potential military conflict within ten years.

"The attention of international politics in the long term will be concentrated on controlling the sources of energy resources in the Middle East, the shelf of the Barents Sea and other parts of the Arctic," the strategy read. "In case of a competitive struggle for resources it is not impossible to discount that it might be resolved by a decision to use military might."

One more hawkish US Russia expert, Ariel Cohen of Washington's conservative Heritage Foundation, wrote, "While paying lip service to international law, Russia's ambitious actions hark back to 19th century statecraft rather than 21st century law-based policy and appear to indicate that the Kremlin believes credible displays of power will settle conflicting territorial claims."

"By comparison, the West's posture toward the Arctic has been irresolute and inadequate," he added.

"Russian Arctic claims are a time bomb," Cohen told me when we spoke in 2009.

By autumn 2010—with a Russian/Norwegian Arctic territorial dispute settled and Vladimir Putin stepping back at least temporarily from tough talk, calling the Arctic a "zone of peace," and with that year's once-scheduled Russian paratroop drop at the North Pole canceled, Russia was cooperating with other Arctic nations in trying to draw up shipping safety and fishery rules for the region.

However, the smooth talk did not necessarily mean Russia had dropped the military planning. Also in 2010, a confidential State Department cable from Prague to Washington—released by WikiLeaks—quoted Russian ambassador to NATO Dmitry Rogozin as saying, "The 21st century will see a fight for resources, and Russia should not be defeated in this fight . . . NATO has sensed where the wind comes from. It comes from the North."

Another State Department cable said that when Russian explorer Artur Chilingarov dropped the titanium flag at the North Pole, "Chilingarov was following orders from the ruling United Russia party."

In other words, the "time bomb" might be less dangerous at that particular moment, but it was still ticking, Ariel Cohen said.

Russian territorial aspirations remained.

Just what would happen if Russia's undersea claims were rejected was anyone's guess.

"Under the Putin regime I don't see things escalating. But under a scenario in which the Putin regime is replaced by extremist nationalists looking for trouble, then even Alaska may come into play," Cohen told me.

"The jury is still out," Cohen said on the phone.

Canada's prime minister Stephen Harper has said, "Use the Arctic or lose it." He's promised to build six new patrol vessels and a state-of-the-art icebreaker. Canada also plans a new winter warfare school and a new Arctic port on the north end of Baffin Island. Canada has announced the creation of a dedicated Arctic military unit. The military installed laser-beam sensors in the Northwest Passage to monitor ships heading through that once ice-blocked route.

In Europe, where Russian bombing runs are also taken seriously, Norway's High North military command lies beneath a mountain near the city of Bodø. When I visited the war room I saw multiscreens tracking ships in the Barents Sea and eyeing the Russian border. Brig. Gen. Per-Egil Rygg, commander of the 13th Air Wing, assured me that relations with Russia were good but then gave me a copy of the *Norwegian Defense Review*, where articles seemed to counter what he'd said.

The opening piece, by Norway's minister of defense, read, "Russia...behaves more boldly and self assertively than before.

This increases the level of tension between Russia and the West...
Russia's attitude toward the west has become tougher, and the war
with Georgia has given the situation a particular edge."

Another article by the chairman of Norway's Parliamentary
Defense Committee warned, "The romantic honeymoon is over
with regards to assessing internal developments in Russia...The
twentieth century was full of...sudden upheavals...Military force
might be resorted to."

A third article by Rear Adm. Trond Grytting read, "Conflicts
may emerge from the High North or may be ignited by conflicts
elsewhere in the world."

The pattern when it came to Arctic pronouncements seemed to
be that leaders in several countries would routinely make aggressive predictions, attack journalists for reporting them, assure the
public that all was peaceful and then start hinting about potential
conflict again.

Keeping that paradox in mind I went back to the State Department and asked about Russia. For instance, in September, a Russian
Il-38 submarine hunter plane armed with torpedoes and bombs
had flown within 170 feet of the USS *Taylor*, off Russia's northern
coast. The *Taylor* had just left the Russian port of Murmansk and
a joint celebration there. The buzzing occurred just as Russia prepared to reassert its claim to the Arctic at the United Nations, and
as Putin prepared a speech on those claims.

Was the buzzing connected to the undersea claims?

I told the State Department official, "I'm trying to figure out
why the Russians do what they do."

"If you figure it out," came the wry answer, "let me know."

Conflict comes in many forms and only one is overtly physical. US
Arctic experts by 2010 were less concerned with actual artillery
being fired in the Arctic but fretted about America falling behind

in the jockeying for influence dubbed by Rudyard Kipling as the "Great Game."

Winners of the Great Game secure riches and block opponents from reaping advantages. Losers, as any history book will show, forfeit influence, power and wealth.

"The person who is present is the one who makes the rules," Mead Treadwell told me in Anchorage.

More State Department cables indicate that Treadwell was not alone in his view. One, slugged "sensitive" and sent to Washington from an oceans policy conference in Greenland in 2007, said that during a conversation about the US's failing to sign the Law of the Sea Treaty, Denmark's foreign minister, Per Stig Møller, told US diplomats, "If you stay out, then the rest of us will have more to carve up in the Arctic."

Another cable said that Arctic Greenland, under Denmark's rule in 2010, was "just one big oil strike away from economic and political independence," that Greenland "might have [oil and gas] reserves to rival Alaska's North Slope," that "our continuing strategic military presence and new high level scientific and political interest in Greenland argue for establishing a small, seasonal American presence in Greenland as soon as possible," and that—with Chevron and Exxon Mobil exploring for oil off Greenland's west coast, the writer of the cable had "recently introduced Home Rule Premier [Hans] Enoksen and Minister of Finance and Foreign Affairs Alega Hammond to some of our top US financial institutions..."

Just being at New York cocktail parties with UN diplomats could give glimpses of the coming strategic importance of the Arctic. At one, at the apartment of a Norwegian deputy consul general, I found myself talking to Jon Erlingur Jonasson, a deputy permanent representative from Iceland.

"The Chinese are all over the place at home," he told me, and the reason was that they were exploring the possibility of financing the remaking of that country into a "Singapore of the North," to handle the flood of expected commercial shipping.

It was clear that most Arctic countries were preparing actively for changes while the US government dithered.

Treadwell posed a worst-case scenario. "Imagine you have some energy-supply disruption around the world. Or all of a sudden we need to move military personnel, need sea lanes open. If that was to happen today, if the Arctic was the ocean we had to pass through, the Russians would have the escort icebreakers. The Russians might decide to shut things down. We've already seen Russia several times in the last five years shut down energy supplies to intimidate another country…"

"The US has to be very careful that we don't give the Arctic to Russia by default," Treadwell warned, echoing the advice of Ariel Cohen at Washington's Heritage Foundation.

"The US needs to elevate its Arctic policy to a national priority," Ariel Cohen urged.

In Washington, DC, that October the admiral drove his late model Saab past a security fence and guard post off Massachusetts Avenue by Reno Road, and past the New Zealand embassy into the grounds of the Naval Observatory—the hilly, heavily wooded compound housing the residence of the vice president of the United States, a Victorian-style mansion. In autumn the oak leaves were dying. A curving two-lane road brought him up to the observatory headquarters building, which occupies the high ground. There he parked in his personal spot, beneath his personal flag that featured two blue stars against a field of white, denoting rear admiral, oceanographer of the US Navy.

Adm. David Titley was a slim, voluble, youthful, white-haired man who used his hands to enthusiastically emphasize points when he spoke. His large office, built in 1893, felt more like a

Smithsonian museum and less like a Pentagon war room. It lay down the hall from the Observatory library, a splendid marble-floored rotunda containing historic works of astronomy and physics that overturned once-popular misconceptions about the natural state of the earth.

It was fitting, considering the admiral's job, that he conducted it a mere two-minute walk from an original *Principia Mathematica*, written in Latin by Sir Isaac Newton, who posited the existence of gravity. And writings of Polish astronomer Copernicus, whose theory—banned at one time by the Vatican—was that the earth rotates around the sun, not the other way around.

"It is almost," the admiral told me, "as if the Arctic has changed its natural state."

He was talking about the melting ice, about the pittance of ice he'd seen from the *Healy*. To Titley, ice-melt questions were far from academic because he'd been given a special job by his boss, Adm. Gary Roughead, chief of US Naval Operations. Titley headed the Navy's Task Force on Climate Change, which had been ordered to make recommendations to the Navy regarding policy, strategy, force structure and investments in the rapidly changing High North.

"Admiral Roughead hired me to think about the Arctic."

Titley sat at the head of his large wooden conference table, eyes flicking to a *National Geographic* map of the Arctic that he'd inserted beneath its glass top, he said, on his first day on the job.

"I've told my boss that by the mid-late century we'll probably see basically ice-free conditions. You'll be able to drive a merchant ship *right across the North Pole*."

This was no academic debate about whether humans had a hand in raising earth's temperatures. It was no argument over global warming theory. The Navy had a pragmatic view. The point was not who or what was responsible. It was that the *Navy* was responsible for US security on the high seas and, to Titley, disappearing ice meant anticipating the strategic implications.

He predicted that the opening of earth's fifth ocean "will not be any different from what our security concerns are in the Pacific or Indian Oceans."

From a historical perspective, this was astounding. He was saying that the Arctic Ocean—which had slaughtered sailors for centuries…with its ice-clogged passages and storms and ice mountains that could smash wooden ships as easily as a shoe pressing down on a matchbox—because of the melting…*would not be any different within a few years, security-wise, from any other ocean.*

All Titley had to do to compare the present and past was to step down the hall to the library or, during a *Healy* trip, to open the science lounge cabinet housing polar books to read volume upon volume testifying to the awesome power of the Arctic's historic rage. There were books dedicated to explorers who froze, ate each other, died of botulism, scurvy, frostbite, exposure, madness or loneliness while seeking a quick sea route—the holy grail of trade—to link Europe and Asia.

He could read about England's lost Franklin expedition, which departed from London with much fanfare in 1845, searching for the Northwest Passage, and took 128 men into the ice on the most modern wooden ships of the era, the *Erebus* and the *Terror*. They were powered by locomotive engines, both by steam and wind. Their propellers could be pulled up from the water if ice closed in. Their hulls were reinforced against pressure. Their living quarters were heated by pipes circulating steam from the ship's boilers. The ships carried canned foods, a new commodity that allegedly would not spoil. Commanded by Sir John Franklin, the expedition was as celebrated in its day as US astronauts first heading for the moon were in the 1960s.

Unlike those astronauts, no one returned. Officers died from botulism poisoning when they ate infected canned foods, considered a special treat denied the enlisted crew. An official investigation would discover that the canner had cut corners, failed to sterilize metal, failed to make sure workers even cleaned their meat-bloodied hands after going to the toilet. Those who survived the tainted food ultimately perished on the ice while trying to man-haul sleds to safety after the ships were abandoned. Their bones, found decades later, showed knife marks on them and human teeth marks. The men had taken to eating human flesh.

Or Titley could read about the British ship *Resolute*, found abandoned and floating in the ice by New England whalers on September 10, 1855. The *Resolute's* crew had been searching for the Franklin expedition when ice trapped the ship a year earlier. The crew abandoned ship, certain that it was about to sink. But the ice eventually released the *Resolute* and, unmanned, it drifted off, red wine still filling glasses sitting in the officers' mess.

The US government returned the ship to England in a goodwill gesture, and the British broke it into timbers that by 2010 sat reconstituted in the White House Oval Office, as Barack Obama's desk. Every US president since Rutherford B. Hayes has used the *Resolute* desk except for Presidents Johnson, Nixon and Ford.

If Titley tired of reading about foreign Arctic expeditions, he could always peruse an account by Charles Brower, in *Fifty Years Below Zero*, of his rescue of some in a US crew who abandoned a whaling ship off Barrow in the summer of 1897.

Brower's account began when, fearing that the *Navarch* would be crushed by ice, he led the men toward shore. They had to hurry because he feared that the solid ice pack was about to split open, and when that happened he wanted to be on the side closer to land. But as the men moved south, the ice they walked over drifted north. They got no closer to land. One man shot himself in despair.

A second fell behind, weighed down by his own clothes, as he had put on everything he owned. Brower recalled the man shouting at his retreating back, "Charlie—don't leave me, Charlie!"

"The next man to go was the blacksmith," Brower wrote. "He hailed from Minnesota. He hung on until the soles of his boots wore through... Soon his feet were cut to pieces..."

Just the titles of Arctic history books usually hint at the fate of the people populating their pages. Leonard Guttridge's *Icebound*, which I'd read on the *Healy*, traces the ill-fated expedition of George Washington DeLong, who believed that the Arctic Ocean would be a great open sea. The USS *Jeannette*—his ship—ended up locked in ice for two years and crushed.

But in 2010 Titley was saying that *the route straight across the top of the planet* could be open by mid-century.

The Navy was faced with many choices.

"It takes ten years to build a ship."

Should the Navy push for icebreaking ships, which might be slower, and expensive, or ice reinforcement for more traditional ships? What kind of weaponry would work under Arctic conditions? What kinds of sensors would be best? Titley had to consider fuel use in the future.

What would happen if the US never signed the Law of the Sea Treaty?

"While the United States has stable relationships with other Arctic nations, the changing environment and competition for resources may contribute to increasing tension, or conversely, provide opportunities for cooperative solutions," said the executive summary of the "Navy Arctic Roadmap."

Titley said, "What we don't want to do is get to the point where it's open and there's massive trade and nobody has thought this through."

And so the admiral spent his days wrestling with logistical questions that, he told me, will be "the drivers of the 21st century."

Arctic changes were even the subject of an actual war game at the Naval War College in Rhode Island soon after we spoke. Within weeks several dozen men and women arrived there from the State Department, Department of Energy, Navy, insurance companies, Exxon Mobil, shippers and the US Coast Guard.

Players concluded that the opening of the Northeast Passage, the Arctic sea route around northern Russia, "could lead to a diminishment of US political and economic power," their final report read.

And that "continued non-ratification of the Law of the Sea Treaty could result in Russia emerging as the dominant power in the region, politically claiming sovereignty of half the Arctic basin, and assuming a lead role concerning Arctic issues. US role in the Arctic could be marginalized if actions, policies and investments fail to keep pace with economic development in the Arctic."

And that when it came to security in the north, "Those with the most influence and leverage may be able to alter the balance as it fits their particular needs."

US passivity in the north could even become a joke as it had at the Naval War College in September of 2009. That conference was attended by representatives from the Navy, Army, Coast Guard, Norwegian government, Russian foreign policy experts, as well as diplomats and maritime analysts.

One day the audience was treated to a talk by US Navy commander James Kraska in which he showed slides depicting the national character of Arctic nations as animals. Russia, with its famed, unpredictable belligerence, was a bear. Canada, with its prickly sensitivity was assigned a porcupine. The United States, he posited, was the traditional proud eagle. But this last slide drew chuckles from an audience that did not feel the image deserved.

That's why the next day a Canadian speaker decided to show his own slide representing US policy in the north.

It turned out to be a sleeping circus elephant.

The audience broke into laughter.

Then the laughter died. To many present, the picture was true.

The big problem was money.

By autumn 2010 nobody in Washington was arguing against the need for more Arctic communication equipment or Coast Guard bases or research in northern waters or against the need to find ways to guard against Arctic oil spills or rescue passengers if a tourist ship hit ice.

But how would the United States pay for these things?

"We're making tough decisions across the board for the budget," Heather Zichal, deputy assistant to President Obama for energy and climate change, told me.

The irony was that Russia *had* more money *because* of profits coming from hydrocarbons from its north. Norway, another Arctic nation, had poured profits from its Arctic and North Sea oil and gas operations into a national pension fund that in 2010 owned 1 percent of all stocks on earth.

But the United States, sitting on its own Arctic hydrocarbons, could not decide on the policy for extracting them—or, in fact, if extraction would be allowed. In the wake of the *Deepwater Horizon* accident, potential policies on the table at the White House had even included the option of refunding the $2.1 billion in lease money to Shell.

Heather Zichal said no one was advocating it, but it was considered. "It was one of the earlier options removed from the table."

Money.

When Admiral Titley gave speeches on the Arctic, he liked to show a slide of Treasury Secretary Tim Geithner. "Because as we deal with all this stuff we're out of money," he told me.

Asked about paying for needs in the Arctic, Heather Zichal said,

"We're trying to focus our priority list when it comes to DOI and Homeland Security through the lens of what we need to do in order to feel comfortable with [oil] production in the Arctic."

It was a catch-22, a round-robin circular dance. You couldn't drill for oil until precautions were in place. Precautions cost money and there wasn't enough of it. Oil income would provide money but you couldn't drill unless precautions were in place.

October turned to November. There was still no word from the Department of the Interior over whether the halt in drilling would be lifted.

For Shell, the clock was ticking. Soon the months would arrive during which Pete Slaiby would need to lease equipment *in case he got the go ahead*. Bills would start skyrocketing.

To try to force a decision in Washington, "we decided to take a gamble," Curtis Smith said. The Anchorage office came up with a plan, but Shell's top lobbyist in Washington didn't trust it. He was afraid it might infuriate decision makers, might backfire against corporate efforts to accomplish anything at all in DC, not just relating to Alaska but to Shell's vast drilling and financial interests throughout the United States.

Shell people in Anchorage wanted to go directly to the public, to launch a national ad campaign to try to push Ken Salazar.

Shell's Washington lobbyists preferred to work quietly, stick with traditional channels—meetings and phone calls.

Arguments broke out between the two offices, the bottom line being "If we don't have an answer by December, we'll have to pull the plug again," Slaiby told Mayor Itta, Alaska's senators, and journalists.

Ken Salazar had a few more weeks to make a decision and then Shell would make it for him. When it came to America's High North, once again, no decision *was* a decision. Indecision by Salazar—staying silent until Shell dropped out—would mean he was making national policy by default.

CHAPTER 9

Caught in the Middle, November

Edward missed his uncle. Over 30 years had passed since George had killed himself, but the memories came often. Itta's wrestling with what to do about offshore oil brought them up.

Sometimes Edward and his old pal Lloyd Nageak ran into each other in town, and one might say, remembering George, "Front is front. Back is back." And they'd both smile.

That lesson had been about "doing things right," and it was a hunting story. The boys were thirteen. George took them out on the tundra, shot a caribou and told them to get their knives out and skin it. Caribou leggings were used to make mukluks, soft waterproof boots, and hunters were supposed to slice the legs in front, then peel off the hide. You cut in front so women could sew the skins correctly. But the boys, starting in back, looked up to see George staring down at them.

"Front is front! Back is back!"

"Edward's uncle was funny," Lloyd said. "Not funny-funny. Funny like Will Rogers. Common sense."

Then there was the lesson about patience. They were hunting

166

again, luckless. Days went by as they boated down a river without seeing caribou; Edward was hungry and impatient and George, finally spotting a few animals, told the boys to wait with the boat. He'd go ashore.

"He said, 'When I get done firing, you boys come.' He said, 'Do not get discouraged for I am with you,'" Itta said.

"We waited for an hour. Two hours. And finally, seven shots. We got on the boat and took off down the river and seven caribou lay there. Nice big ones. 'Didn't I tell you I am with you?'" Uncle George said.

But that lesson—*those exact words*—turned out to be far, far bigger because the second part—about faith—didn't come until years after George shot himself, when Edward struggled through his dark time in San Francisco. "I was almost unbelieving in that period of drugs or alcohol, wondering what life is about. I was reading the Bible. I saw the line *'Do not get discouraged for I am with you.'* And it hit me. The epiphany. He knew there was a higher power. My uncle did not say things directly. He'd left me a clue."

Those words reached deep inside and began pulling Itta away from the dark. It was as if George had known the struggle would take place.

George was always a part of his nephew, and as November began and Edward awaited a decision from the Department of the Interior, he often recalled his uncle's wisdom about land. George had meant onshore land but in 2010 the words applied to the seabed too.

"My uncle told me when the land-claims issues bubbled up that our world would change. He said, 'We won't have this land anymore. Someone will say, *I own it*, and we need to be ready for that.' And later when the call came to register and join the Arctic Slope Regional Corporation my uncle said our life *had* changed forever. He was a learned man. He knew the Constitution. He said we were getting shortchanged on land and never signed up. I almost never did either."

Now, in 2010, outsiders were fighting over who owned the seabed and whether to drill on it. The *Healy* had been out at sea making maps, taking samples. The State Department was readying a claim. *We own it.*

"I cannot help but feel that what is happening now is similar to what my uncle dealt with. A life changer. A culture changer. I cannot help but feel the changes were tied to my uncle's demise."

The images haunted Edward: his uncle reeling from shocks to Eskimo life, seeking solace in alcohol. George so desperate that he'd drink anything that contained it, even cleaning fluid. His mind altering, his character darkening. Edward so distraught after the shooting that he could not attend the funeral, finding out later that George had loved him so profoundly that he had left him his house.

Itta staggered away from that tragedy having learned about the need to prepare for change's dangers and the consequences of refusal to adjust. And these lessons informed his days in 2010 as he tried to coax, finesse, threaten or negotiate with Shell and federal officials into adopting policies he felt would protect the seas in which whales lived and in which oil exploration would one day inevitably—in his mind—take place.

"My uncle had faith in me. He told me, 'You will be able to do things that I'm unable to do.'"

Itta wished that George were there. Or Eben Hopson. The leaders who had come before him. "It's a lonely time here. I have nobody to really talk with."

Sometimes small surprises bolstered him. One time he was with his father and one of his dad's elder friends. The elder told Edward, "You guys [Itta's generation] know so much more than we do. We're scared to talk to you."

It had not occurred to Edward that his wrestling between

worlds—Western and Iñupiat ones—might be valued by his community. No one had told him that he was *special* that way.

"I resolved that I had value. That the melding of two worlds made me what I am."

Edward was always upbeat and jovial around people when I was there—telling jokes, brightening the room. The bad memories only came out in private and only if I asked about them.

By November, Itta was still in the middle, trying to keep other native groups like the Alaska Eskimo Whaling Commission from joining environmentalists seeking to stop Shell in court. He thought more would be accomplished with a seat at the table. He feared, based on experience, that once lawsuits started all parties would be prohibited by lawyers from talking. This had happened when the borough had sued to stop the Department of the Interior before. Edward hated the memories of not being able to try to work things out personally with people he'd easily talked to before.

"Once you sue, it becomes all or nothing."

And in an all-or-nothing situation, he believed, in the end, the North Slope would lose out.

He also believed the national environmental groups—given a free hand—would stop all drilling and block off immense areas of the North Slope from any kind of development in their efforts to save animals.

"They'd create a new endangered species. Us."

"Problems happen when outside people want to help us," Richard Glenn said in one meeting I attended between Itta, ASRC officials, and Sen. Mark Begich in Washington, while Itta—who had asked Richard to speak—nodded. "People want to put us in a snow globe. Nothing new goes in or out. My father used to say that sustainable development means that no baby can be born until someone dies."

Itta was pushing in Alaska's Capitol in Juneau that month to establish a state commission to manage offshore issues. Pete Slaiby hated the idea of more regulation.

Itta was also about to sign the science agreement with Shell to cooperatively conduct research on issues relating to the continental shelf, but when he wrote Secretary Salazar in late November he made no mention of this.

In Itta's letter he called Shell's 2011 Beaufort Sea exploration plan "clearly an improvement over prior plans" as it was for one year only, included more blackout dates to protect the bowhead whale hunt and proposed to haul away discharges that former plans proposed to dump into the ocean. While some of these changes had been forced on Shell "by more than four years of discord and legal challenges," Itta wrote, and "adopted by BOEMRE and industry because of court decisions and community concerns," others, like the backhauling of some discharge, reflected a voluntary commitment on Shell's part "to move toward a program with fewer impacts on our communities."

Itta suggested that the whole struggle might have been avoided "if there were a better system for mainstreaming our ideas and concerns."

He added that although he opposed offshore development in principle, the shrinking oil flow through the Alaska pipeline, the federal policy banning onshore drilling in ANWR, and the determination of industry and government to explore offshore in the Arctic made it "more difficult for me to stand firm against any and all offshore activity." But that did not "lessen my responsibility to defend against any offshore activity that threatens the health of the ocean."

Accordingly, Itta pushed for Salazar to make Shell's voluntary changes into requirements for all companies seeking to drill off the North Slope. "It is difficult to imagine agencies will enforce standards that are not part of a permit." He also cautioned against approving Shell's plan before other permits that the company sought, like the EPA air permit, were granted.

He wasn't suing but he sure wasn't allied with Shell.

Still, the letter implied more opposition to the 2011 plan than I'd felt in him privately. Wondering how much of this was maneuvering—a need to keep pushing, to never relax—I called him in late November 2010 to ask.

"Mr. Mayor, if Secretary Salazar called you today and said, 'It's up to you. Should we let Shell drill *one well* in the Beaufort Sea in 2011 or not? *Have they modified the original plan enough for that much?'* what would you say?"

There was a thoughtful silence on the line. I added, "I know you feel responsibility. I know how much you want to balance revenue with protection. I know you have reservations. But bottom line. Yes or no?"

Itta sighed, "It's funny you ask this. My staff and I just talked about it."

He paused.

"I'd say yes," he said.

I also wondered what the Coast Guard thought about Shell's plans for handling oil spills, so I asked Adm. Chris Colvin about it. He headed all operations in Alaska.

"I'm comfortable with them drilling an exploratory well in open water when you have light 24 hours a day," he said.

Colvin said that Shell had "gone overboard" in making safety arrangements. He was fine with the exploratory well being drilled in ice-free summer. But he questioned whether any existing technology at all could clean up oil spilled in ice. If it ever came to the next stage, a producing well that would operate year round in ice, his feelings were different.

"They'll need a game changer," he said.

——

By that November, Shell's goal—to Pete Slaiby—remained modest. He wanted to get a foot in the door. Here's what the exploratory scenario would look like the following year if he got permission to sink a single well.

Early in July 2011, the 511-foot-long drillship *Noble Discoverer*— a converted logging ship from Southeast Asia—leased by Shell from Swiss owners, would leave Dutch Harbor in the Aleutian Islands, squeeze past Russia in the foggy Bering Strait, enter the Chukchi Sea, and, heading east over roughly the same route that the *Healy* had taken, reach a drill spot in the Beaufort Sea after an eighteen-day transit. It would be roughly fifteen miles from shore and 70 miles from the nearest village, Kaktovik.

On board and given full run would be an inspector from the Department of the Interior, 24 hours a day, able to stop operations if he or she saw a problem. The inspector would have satellite phone access to Washington so there would be no lag time if a problem arose.

When federal regulators had moaned about difficulties they faced in the Arctic, the great travel distances, the question of where inspectors could stay, Pete Slaiby had made Shell's offer clear. *Hey! Our ship is yours! We're not hiding things! Inspectors don't live on rigs 24 hours a day in the Gulf of Mexico but you can do it here. Just let us begin drilling, okay?*

Also on board would be marine-mammal observers on the bridge, the privately hired watchdogs—kids just out of college, retirees, hardy outdoor types, young people trying to figure out what to do next in life—whose job it is to spot whales, seals or walrus in the path of an oncoming ship. By law, they can order the captain to stop, slow down or veer away if the ship gets too close.

A company that fails to heed a warning faces fines and even project shutdowns.

The drillship couldn't even enter the Chukchi unless observers were aboard.

Slaiby made that part clear in public hearings too.

All offshore activities would be monitored by native observers manning communication centers in North Slope villages, where—as in a war room—ship movements would be marked on maps. Elders would be consulted daily as to the location of mammals and would get their information from local hunters.

Other changes to Shell plans had been made due to lessons learned from the *Deepwater Horizon* disaster, some ordered by the Department of the Interior. The *Discoverer* would carry a "capping system" to be lowered onto and plug a ruptured wellhead (where oil comes out of the ground) in the event of a blowout. A barge nearby would carry a containment system—a huge box-like vacuum designed to suck up spilling oil and funnel it to the barge for storage. The drillship would have additional support vessels nearby; an icebreaking tug and a couple of smaller oil-spill-response boats.

The rig's blowout preventer had been upgraded.

Pete loved oil wells the way a mechanic loves a brand new Jaguar engine or a pilot admires a Boeing 767 lifting off the ground. A drillship easing into the Arctic, to him, would mean cheaper heat for homes in Boston in winter and more dollars to pay for roads in Alabama, bridges in Iowa, schools or teachers in Vermont, California and Florida.

And safety? When critics accused any oil company—not just Shell—of sacrificing safety for profit, Pete would do a fast burn. He'd see his own ghosts, flashing back to England and a day when, in 2002, coming home from a 25 mile copter trip to North Sea gas platforms, he'd sat down to a delicious spaghetti dinner with

Rejani when the phone rang and a caller said, "*Victor X-Ray* just went down."

Pete and Rejani stared at each other in horror, knowing that six people had just died in a copter crash—the same copter that had carried Pete home. It could have been him on the bottom. The dead included parents, men he worked with, guys he'd hung out with the night before and one worker who had even thanked Pete for his job.

"They were friends."

Pete changed after that, Rejani said, going quiet and depressed for months. He went to the funerals, comforted the widows. The accident had happened—he learned—because a chopper blade—refurbished years earlier after being damaged by a lightning strike, and rebuilt under FAA standards—had cracked anyway, sending 12,000 pounds of unbalanced copter to the bottom of the North Sea.

"There's not a July 16th that goes by that I don't think about it. Shell doesn't pay me enough to gamble with lives."

Pete did not usually lose his temper, but if someone questioned his commitment to safety you could see him get mad in his quiet way, his face going still, his eyes losing light, the ex–bar brawler trigger in view. One time in a public meeting Curtis watched Slaiby almost lose it. There was this guy in the audience, prototype of "a guy in every audience," Curtis said...maybe well intentioned but ignorant, as far as Curtis saw it—ex-hippie maybe, gut liberal, a barefoot guy in Crocs going on about Shell's ignoring safety, Shell's lying about their Arctic experience, repeating some lie snatched off the Internet—and Pete leaned over the stage as Curtis went rapt, wondering if he was about to see the bar fighter come out, hearing Pete hiss, just loud enough, "Listen buddy, I don't know *what the fuck* you're talking about, and neither do you."

Once.

The rest of the time, in testimony or village hearings the engineer remained in control. Harry Brower Jr.—by 2010 head of the Alaska Eskimo Whaling Commission—recalled a story about Pete's response to a challenge once. It had happened when Slaiby had first come to Alaska and was attending a meeting with whalers in the Cook Hotel.

Pete told the Eskimos at that time that although Shell planned to discharge cuttings—shavings from drilled rock—into the sea, the process would be safe. Harry got mad because "our animals that we depend on for food were going to be swimming through all this stuff." Finally Harry snapped. "We were getting ready to be served lunch. I said, 'Mr. Slaiby, why not take that tray of food in front of you and run it by your drill rig through that discharge you were just talking about. Would *you* eat that?'

"Pete Slaiby didn't say anything. He asked to be excused. He left the meeting and came back ten minutes later and his face had changed and he said, 'Mr. Brower, I have been told to answer your question. My answer is, no, I would not eat that food.'"

Harry responding, "So is it good for my people to eat it?"

"That was the last time Shell invited me anywhere. But now—in 2010—they've agreed to collect all that waste and transport it away."

By November 2010, Pete was claiming oil spilled in ice conditions could sometimes be *easier* to clean up because ice kept it from spreading. Sometimes. Pete would say that the North Sea is a tough place in which to clean up oil too.

Pete had an engineer's confidence in technology. The belief that he would handle any problem that came up.

Pete believing—continuing Shell's wish list for a smooth 2011—that reaching the drill site, the *Noble Discoverer* would set down eight anchors to keep the vessel stable while working.

During the next 30 days of operation, the *Discoverer* would, as its clean-air permit would mandate, release only safe levels of pollutants, Pete said.

Pete explained to audiences that in the unlikely event of an oil spill, the *Discoverer* would not face the problem alone. Another Arctic-class drillship, the *Kulluk*, would be present to act as an emergency drill rig if necessary. Shell had leased backup ships and would have local teams on site to operate cleanup booms, skimmers, and oil burning devices as well as a backup remote controlled blowout preventer in case the one on the *Discoverer* failed.

The skimmers—a sort of giant rolling mop system—could soak up oil and deposit it in a storage tanker able to hold 500,000 barrels, which would be on site within four hours of a spill. It could not stay closer to the drillship during normal operations because if it did, Shell would exceed the EPA's allowable air emissions at the site.

Back in Anchorage, Geoff Merrill—Slaiby's emergency-response manager—had been buying cleanup equipment including the "NOFI Current Buster," which was basically a V-shaped inflatable boom. Towed by two boats into an oil spill, the giant raft would corral oil inside. He'd also purchased "First Mate Compact Thermal Night Vision Cameras," guaranteed to give an oil cleanup crew the ability to "see people, vessels and navigation hazards in total darkness."

So Pete wasn't in his own mind some uncaring corporate robot ready to blow up Mother Nature to help win a raise. He *was*, at home, a rock-solid Republican who was so angry about what he considered big government's blocking progress that he'd even considered running for office.

At home, baby Teddy in his arms, he'd look at the child and think about the boy's future and that the country needed to *produce* things to remain healthy by the time Teddy became an adult.

Asked that fall by a reporter, "If Shell is taking new oil-spill precautions, doesn't that mean the old ones were insufficient?" Slaiby retorted, "Then you'd never do anything. You'd never fly on a plane because there would always be a safer one later on."

"If we don't get movement by December 2010, we won't be able to drill in 2011," Slaiby said.

In late November, Slaiby boarded a plane to Washington. This would probably be his last trip there before Secretary Salazar either ended the drill suspension or Shell pulled the plug on plans. By the end of the month there would be an answer one way or the other.

"There are no more plays like this in the world," said Susan Childs, Shell's permitting person in Alaska. "You don't walk away from a play like this. We won't go down without a fight."

In Washington, Sen. Mark Begich was trying to get an answer from Salazar too. Begich is a youthful, articulate, amiable man who tells funny jokes and gets passionate about issues and displays a bumper sticker in his office that says VEGETARIAN. ESKIMO WORD FOR POOR HUNTER.

He'd been calling Salazar at the office and at home.

Begich even went to the White House for help. After President Obama named Pete Rouse—onetime Alaska resident—as chief of staff, Begich dropped by 1600 Pennsylvania Avenue to make the pitch.

"I said congratulations. Then we got down to business."

Begich talked about the national need for energy but also talked politics. He'd voted in the Senate for President Obama's national health care bill, which was, to put it mildly, reviled in Alaska. He was vulnerable to losing office. Begich asked Rouse to go to bat for Shell. The argument was, basically, *I voted for health care. Give me something back. If you want to keep a Democrat in office in Alaska, help me out.*

"I'm not bashful about leveraging," Begich said.

Rouse said he'd pass along the senator's concerns, which

Salazar knew anyway since he was regularly getting phone calls from Begich. Rouse said he wouldn't push Salazar one way or the other.

Rouse told me, "The administration is not predisposed against the Shell lease. The concern is that the outer continental shelf in Alaska is a separate environment where we don't know much."

Also at the White House, Heather Zichal said that the secretary was making "science-based decisions. In light of what happened and lessons learned from the *Deepwater Horizon* spill, we're applying those lessons and looking for new opportunities in the Arctic Circle through that lens."

Rouse and Zichal said that Obama did not weigh in.

"This administration reflects the independent decision-making authority of the agencies," Zichal said. "We don't meddle in the permitting process."

For months, Salazar had been holding public meetings in Alaskan villages and talking with oil executives and representatives of national environmental organizations. He'd been following the Deepwater Horizon Commission investigation and its criticism of the oil industry for failing to adapt to new safety concerns as they moved into challenging areas for drilling offshore.

Begich told Salazar, sure, that was fine, but, "At some point you have to say it's okay for them to go for their permit. You can't keep doing the same old 'We'll study it.' "

Underdogs.

Most people don't think of global oil companies—with their access to world leaders and armies of lobbyists—as underdogs, but by November 2010, Pete Slaiby felt like one.

The process had become a never-ending do over game. Every decision was made, remade, challenged, rechallenged and then litigated.

At the EPA, for instance, the clean-air permit *had been granted to Shell* but was being appealed through an inside-the-agency process. Environmental groups like Oceana allied with the Alaska Eskimo Whaling Commission and Native Village of Point Hope had requested the three-person panel review.

If the permit was approved, "it will probably be challenged in court," said environmentalist lawyer Peter Van Tuyn.

No one was in control. Individual parts of the process made sense but when they came together they were contradictory, redundant and complex to the point that there was no way to navigate the system.

In Washington, agencies did not talk to each other.

The broken system had existed under both Republican and Democratic administrations. The pro-oil Bush White House had pushed to open new areas to oil companies, and later courts ruled that shortcuts violated laws. Under President Obama, after the Gulf disaster—agency administrators slowed things down while conducting reviews.

Add it all up and Mead Treadwell, who as of November was Alaska's new lieutenant governor, said, "If you want to get permission to drill offshore here it's like the most complicated video game you've ever seen where every piece of floor could be a trapdoor, everything above you could drop down and eat you in a moment."

The day before Pete got to Washington, I rode the Metro to meet with Deputy Secretary of the Interior David Hayes, part of Salazar's decision-making team. Reading material in the car provided commuters a view of the battle for the Arctic.

NATURAL GAS! GET ON BOARD! WHAT ARE WE WAITING FOR? screamed the banner headline on ads plastering the car, sponsored by the American Clean Skies Foundation.

I also caught a glimpse of one man reading an ad in the *Washington Post*, part of a media blitz designed to swing public opinion in Shell's direction.

The ad showed what appeared to be a super tugboat on steroids—a large, powerful-looking ship at anchor beneath snow-dusted volcanic peaks and misty clouds. Blue hull. White super-structure. Orange winch on back. The northern landscape seemed formidable, but no problem for the vessel. The two hard-hatted men standing on the aft deck struck a viewer as competent and attentive.

LET'S BE MORE PREPARED THAN EVER, the headline read.

I was looking at the "all-purpose-built, ice class, spill response vessel *Nanuq*," part of Shell's Arctic response fleet, which would be on the drill site scene.

Over the commuter's shoulder I read, "Shell is dedicated to preventing and preparing for any safety challenges we may face. To support our plans to drill two 30-day exploration wells in the shallow waters off the coast of Alaska in 2011, we have created an unprecedented spill response...Of course, our goal is to make sure we never have to use it."

It was fairly normal in Washington for political battles to spill over onto the Metro. That's because the Washington Metro system has some of the most interesting subway ads in the world. Unlike ads in, say, New York subways, which target an audience suffering from zits, bunions or a desire to sue people over botched surgeries, sexual harassment or traffic mishaps, Washington ads are aimed at all those riders in belted raincoats and sensible shoes who help spend trillions of government dollars each day when they get to work.

Underground in DC you don't see posters showing lepre-chauns wearing boxing gloves (OUR LAW FIRM FIGHTS FOR KIDS!) but sparkling new missile-guidance or satellite systems to help blow up the nation's enemies or chat with friends in Zaire; Kuwait or Barrow, Alaska.

"If you're a defense contractor, you're going to target the Pentagon station," Brian Malnak, Shell's Washington-based chief lobbyist, said. "If you're international, you'll target Foggy Bottom. If you want the Hill, it's Capitol South. The trains travel very specific routes."

Shell's ads, running in major newspapers and magazines that month, had been the subject of arguments between Peter Scott, Slaiby's communication chief, and Brian Malnak, a politically savvy, fast talking ex–New Jerseyite whose K Street office sat within walking distance of the *Washington Post* and Loews Madison Hotel, where Slaiby stayed in DC.

Malnak's job was to follow developments coming out of Washington that could affect any Shell operation worldwide, and he'd been concerned that the *Nanuq* ad and a LET'S CREATE 35,000 JOBS, ONE STEP AT A TIME ad, also running nationally, might alienate politicians who dealt with many other oil issues, not just Alaska.

Malnak was convinced to go along.

Leaving the train at the Farragut West stop, I hurried over another enormous ad on the floor, this one from the Alaska Wilderness League, showing a cute baby caribou.

ARCTIC WILDLIFE MONUMENT. A LEGACY WORTH PROTECTING, the ad said.

The Department of the Interior Building is massive, larger than any building in Barrow, situated three blocks south of the White House on Eighteenth Street. Inside headquarters work the stewards of America's natural wonders. Once dubbed the "world's finest office building" because it had central air conditioning when it was completed in December 1936, it resembles an old block-sized Soviet-era granite lump.

Hayes's sunny office was brightened by colorful oriental carpets, which, while aesthetically pleasing, could not obliterate the sense of apparatchiks working outside in offices lining long empty halls. But Hayes was no lifelong bureaucrat. A cordial, insightful man, he was a former partner in the national law firm of Latham & Watkins, and a former deputy secretary of interior under President Bill Clinton.

Hayes's views on Big Oil were known at Shell. He'd criticized Bush-era policies in a 2002 Progressive Policy Institute paper, writing, "A short term emphasis on increasing domestic oil and gas production from public lands may compromise important environmental laws and values, damaging lands held in public trust for future generations in return for limited short-term supply gains."

And that the "strong focus on increased production...to the apparent exclusion of—or significant discounting of—competing environmental impacts, puts our nation at the risk of repeating the boom/bust energy cycle by recklessly developing energy plays that provide short term benefits, but which leave lasting negative impacts on public lands that we are holding in trust for future generations."

Hayes told me that Salazar would make the decision on the Arctic "within the next few weeks" as part of a general appraisal of future drilling in the region.

"We feel an obligation to honor leases in the Arctic...that the prior administration granted," he said. "That companies like Shell paid a lot of money for...Are we comfortable that exploratory drilling can be done safely? That's a question we have a lot of expertise on, a lot more than we had six months ago. Can this next phase of exploratory drilling, in a limited time frame and in a limited way, be done safely or not? That's the key."

Hayes said that another question was whether the Arctic would be opened by the department to more drilling in an upcoming 2012–2017 plan to be drawn up soon.

"But we don't have to answer that question now."

Was the big Shell advertising campaign having any effect on Salazar's decision making?

"We're pretty immune to that stuff at this point. We're trying to call balls and strikes in a timely manner."

Hayes acknowledged that "Shell and other companies have... upgraded their containment systems based on the Macondo well." I asked if Salazar was satisfied with the precautions.

"We're not making that decision right now."

Were Mayor Itta's views being considered?

Native Alaskans were important in the process.

"What *is* your impression of Mayor Itta's view? Do you think that if it were up to him he'd allow Shell to drill a single exploratory well or not?"

Hayes didn't know for sure.

The Department of the Interior supervises more than half of Alaska, and Hayes was "trying to take a hard look at what makes sense there," he said. The Bush administration had "made an eight-year effort to open ANWR for drilling and it was unsuccessful." And had "considered but ultimately backed off leasing some of the most sensitive areas of the National Petroleum Reserve. We're glad about that as well.

"There are some areas that just should not be drilled."

Hayes ended the meeting by saying, "We want to be sustainable for future generations. So folks will look back on this period and say, 'Those people were thinking about the right values when they made decisions.'"

The decision came the next day, December 1.

Washington temperatures were in the 50s, tropical for the Anchorage crew. After breakfast at the Madison, the Shell group met at Brian Malnak's office to plan the day's lobbying. Pete would be interviewed at the *Washington Post* and *Politico* and then drivers would take the party to meetings on Capitol Hill and at the State Department.

"We want to start developing a relationship with the State Department. The Trans-Alaska Pipeline is clearly a strategic national asset," Slaiby's political adviser Cam Toohey said.

Also, "The State Department is starting to get a lot of inquiries from foreign governments asking what we're doing about leases in the Arctic."

Pete was also scheduled to meet with Heather Zichal at the White House and with Secretary Salazar. But things changed at the 8 a.m. strategy meeting when the group began getting e-mails from Houston.

"We learned that Salazar was going to make a big announcement," Cam said. "It might include the Arctic."

By 10 Pete was at the *Washington Post*, where the reporter interviewing him knew that an announcement would take place but did not know what it would be.

By 11 he was in a Chevy Suburban on his way to talk to Alaska's sole congressman, Don Young. As the Suburban made its way along Independence Avenue past the National Museum of the American Indian, nobody was looking out the window. Everyone was on their BlackBerrys.

"When Pete gets nervous, he likes to move around but you can't move around in a car," Cam Toohey said.

The only sound was the clicking of fingers on buttons. Pete tried to call Shell president Marvin Odum in Houston but he was not getting through.

Cam insisted that the announcement would be good.

"If you want to let something out that's crappy, you put it out on Friday at 4 p.m., so it misses the news cycles. This is midweek. It will probably be good," he said.

Still, he was guessing. Then a call came from Senator Begich's office. The news would be positive.

Soon after that, Salazar announced in a press conference that the Bureau of Ocean Energy Management, Regulation and Enforcement would continue to honor existing oil and gas leases in the Chukchi and Beaufort Seas but do so with the "utmost caution."

The suspension was over, but this did not mean that Shell would *get* permits, just that "we are reviewing and processing Shell's application to drill a single exploratory well in the Beaufort Sea," BOEMRE director Michael Bromwich said. A public comment period would now commence during which the agency invited opinions on Shell's plan. It would end on December 22.

Bromwich added that the ultimate decision would "not be constrained by any deadlines. We understand Shell needs a decision, and when we complete the review and analysis we will be in a position to make our decision."

Translation: Shell was in the same holding pattern in the Beaufort that they had been in on March 31 the year before.

Reached on the North Slope, Mayor Itta was cautious. "The process should go forward, but let's not forget there are still outstanding issues," he told *Petroleum News*, an industry trade journal.

Asked whether there was enough time left for the agency to process Shell's application by the start of the 2011 drill season, Salazar told me there was, but there remained "issues at EPA."

The environmentalists began celebrating at 6:30, after the sun went down on December 2. Cabs pulled up to the Newseum, Washington's media museum, on Pennsylvania Avenue. Valet parking attendants whisked away private vehicles.

The guests were in Washington to mark the 50-year anniversary of the Arctic National Wildlife Refuge, which sat above the same geological formations that contained Shell's offshore lease areas in the Beaufort Sea.

David Hayes joined the crowd streaming in. So did New York ex-congressman Bob Mrazek, who had been designated a "hero of the Arctic" by tonight's hosts at the Alaska Wilderness League. Also in attendance were Anchorage lawyer Peter Van Tuyn, and Robert Thompson, the Iñupiat president of REDOIL, a North Slope native group opposing all offshore extraction. Former president Jimmy Carter was scheduled to give a speech by remote. This was a happy night in a venue that, considering the grim photo exhibits and headlines displayed—battle shots, terrorism news, air crashes—commemorated conflict, without which the press would cease to exist.

Present in the crowd were several people who actively fought Shell's offshore plans.

"Ask Pete Slaiby how they're going to clean up oil under ice?" Robert Thompson demanded as we waited in line. "It's impossible."

"Ask him how they can clean up oil if it spills on September 30th, just as ice moves in."

"Ask how they could clean up oil even in summer if there are big waves. Ask how come Shell is so sure that the wells up there would be low pressure."

As for the science agreement between Mayor Itta and Shell, Thompson said, "It's going to be used to support Shell, so why did the mayor do it, eh?" He suggested financial reasons. The mayor was a stockholder in the Arctic Slope Regional Corporation, which worked with Shell.

Thompson, a tireless fighter, was a blocky, gray-haired wilderness guide and hunter from the village of Kaktovik. Beside him was old friend Peter Matthiessen, revered naturalist author, whose writings on the North Slope's endangered wildlife had stirred emotions around the world.

The bar was upstairs. The atrium had been turned into a dining area. Round tables were covered with linen and china. The menu included salmon and baked Alaska. Old friendships were renewed as activists sipped wine and munched hors d'oeuvres beneath a brightly lit electronic headline streaming a nonstop list of "heroes of the Arctic"—enviro-warriors who had helped create or sustain the ANWR wilderness by the Arctic Seas.

"Shell is leading us down a primrose path when it comes to offshore in the Arctic," Peter Van Tuyn told me, glass in hand. He was a curly haired, bearded advocate whose passion matched Mayor Itta's and Pete Slaiby's. "Pete Slaiby is a mouthpiece. He stands there with an open, honest gaze, looking into the camera, and says, 'Don't worry, we can clean up a spill if it occurs.' He's wrong, just like all those promises oil companies made in the Gulf of Mexico."

Van Tuyn told a story. He said he'd gone on an oil-spill-cleanup test sponsored by BP near Prudhoe Bay in the late 1990s. All the high-tech descriptions of the process, he said, boiled down to no more than a gigantic mechanized mop/squeegee system. A hydraulic arm swept over water that watchers were supposed to imagine covered with oil, and enormous rollers soaked up the "oil." Then— no different from a husband wringing a mop soaked with toilet overflow—the imaginary oil was squeezed out over a barge.

Some futuristic system, and it didn't even work, said Van Tuyn, who worked closely with attorneys at Earth Justice and represented the Native Village of Point Hope and Alaska Wilderness League against Shell.

The boat got stuck, he said. The enormous mops got tangled. Van Tuyn was astonished afterward when BP officials announced the test a success.

"When we challenged that," he said, "the BP oil-spill contractor responded that the exercise had been designed to test a hypothesis. The hypothesis had been proven wrong, so, as a scientific matter," said Van Tuyn, shaking his head, "since they had an answer, they said the test was a success."

These days, "I don't see a scenario where Shell gets permission to drill and the conservation community won't sue," he said.

The basis of the suit would depend on "how the Department of the Interior makes a decision," but whatever the basis, there would be a suit.

"The US Geological Survey report on the state of Arctic science isn't even due out until April 2011, so how can DOI let Shell drill? The *Deepwater* oil spill commission report won't even come out until January," he added.

Van Tuyn was pleased that laws provided several avenues whereby suits could be brought to stop drilling. The Endangered Species Act, under which polar bears had been ruled as threatened, might come into play in 2011, he said, especially if oil company activities were scheduled to take place in bear habitat, which, since it is ice, could be argued to be anywhere you look on the coastal North Slope.

Shell, he said, was "short-sighted and stupid because it thought the courts were going to be as easy for it to roll through as the Bush administration. Shell knew it could get anything, no matter how lame, through the Bush administration."

Shell, he added, had "committed fundamental blunders and lost hundreds of millions of dollars as a result . . . I don't think we ought to pay for their mistakes." The blunders in his opinion included pushing to get leases during the Bush years, inadequate science, the whole premature effort.

"Shell would like to have certainty, but we can't let that drive us."

Van Tuyn thought the process by which decisions were made in courts worked just fine. They did what they were supposed to do, enforced laws passed by Congress.

Van Tuyn had experience on both ends of these laws as, early on, he worked as a lawyer for the Justice Department during the George Herbert Walker Bush administration, where, "I did black hat stuff as well as white hat stuff," he said.

Downstairs, during speeches, the screen showed stunning shots of ANWR's rivers, tundra and caribou migrations. An American Serengeti. Someone called for Van Tuyn to take a bow for protecting the Arctic. To rousing applause, he did.

Nobody from Shell or Mayor Itta's office attended the affair, during which former president Dwight Eisenhower received the Arctic Legacy Award, for establishing the Arctic National Wildlife Refuge and Jimmy Carter received the Arctic Stewardship Award, for expanding that wilderness.

"As long as President Obama is in the White House, as long as Secretary Salazar is at the Department of the Interior, there will be no drilling in the Arctic!" said a speaker.

The applause went on for a long time. It was a happy night. To people present, even one well in the offshore Arctic would mean the opening of a Pandora's box, a door that, once opened, would never close.

That December, elsewhere in the Arctic, the British company Cairn announced that its drillers had discovered two types of oil in the Baffin Bay basin off Greenland. The Canadian National Energy Board approved construction of a $16.2 billion Mackenzie Valley pipeline project to carry land-based natural gas from the Beaufort Sea area south. The pipeline could eventually carry offshore hydrocarbons as well.

In Russia, Rosneft, the country's largest state-owned oil firm, announced that it would start producing natural gas from the Kharampurskoye gas field in the Arctic; and Gazprom Neft, the fifth largest oil company, prepared bids for leases beneath the Kara and Pechora Seas.

In Washington, as the party went on in the Newseum, Pete Slaiby was across town at the Palm Restaurant, allowing himself a small celebration but mostly a working dinner. Over fish and steak, or cereal and eggs next morning at the Madison, he, Curtis, Brian Malnak and Cam Toohey were mapping out what needed to be done if a single well was to be drilled in the offshore US Arctic that year.

"In a way it would have been better if we'd been turned down," Brian Malnak said. "At least we'd know the answer."

After so many setbacks, there was a Pavlovian expectation that once again, good news would turn to bad. But the mood rose later that day when Slaiby walked into the Department of the Interior and met with, for the first time in the same room, Michael Bromwich, newly appointed head of BOEMRE, and Monica Medina, deputy undersecretary at NOAA and US commissioner to the International Whaling Commission.

"They committed to working together on the permits."

So began joint meetings—Shell engineers and NOAA and BOEMRE staffers—30 to 40 people at a time jointly poring over permit applications as the feds decided on whether Shell would get permission to drill.

Cam Toohey said, "For a long time we'd been asking for one-stop shopping."

Pete liked the cooperation but still did not know who would make the final decisions.

"That's the maddening thing about this process."

He sensed movement but said, despite it, "Do I feel like there's anything really different in the hearts of the regulators? No, I don't."

CHAPTER 10

The Frustration Mounts, December

It was called a controlled crash landing. The pilot of the chartered PenAir Saab 340 circled warily above clouds obscuring the Aleutian island of Unalaska, 800 miles southwest of Anchorage. At a small opening the little turboprop plunged through, past volcanic peaks. The wheels smacked down amid driving snow slanting across the short runway. It was December 9.

As the jet came to a stop, Pete Slaiby and a handful of Shell staffers and their guests—eighteen Iñupiat and Yupik leaders from coastal and Bering area villages—let go of the armrests in relief and looked out at Dutch Harbor, Unalaska's port and home to the popular TV series *Deadliest Catch*, about Alaska's king crab fishermen. It was also the last port visited each summer by the *Healy* before heading for the Arctic, the second largest seafood port in the United States after New Bedford, Massachusetts, and the possible future Singapore of America's High North, according to Scott Borgerson, Oceans Fellow at the Council on Foreign Relations.

Shell's guests had come for a close-up look at equipment designed to handle Arctic oil spills, one of the greatest potential nightmares in North Slope villages. The vision of thousands of barrels of crude oil gushing from a ruptured drill pipe or holed tanker, or from an iceberg-gouged underwater pipeline struck terror into the hearts of whale hunters and their families and anyone who ate seals or walrus or fish or simply appreciated the pristine views, the ice and seas of the High North.

In 2010, some spilled oil from the *Exxon Valdez*, the tanker ship that ran aground in 1989 in Prince William Sound's Bligh Reef, was still found in spots in Alaska.

To allay fears, Mayor Itta had asked that cleanup tests be conducted in the Arctic, but the EPA prohibited them. In other words, everybody argued over whether it was possible to clean up oil in northern conditions, but federal rules prohibited the tests that would answer the question, the same question that kept government agencies from making better decisions. It was another catch-22.

Accordingly, the only tests done had been carried out in Europe, funded by a consortium of oil companies, Shell among them, in 2009. The tests occurred in the Norwegian part of the Barents Sea east of the Svalbard archipelago, in ice cover ranging between 50 percent and 95 percent.

Partners in the program included the Swedish Coast Guard, Norwegian Clean Seas Association for Operating Companies, University of Alaska–Fairbanks, University of Rhode Island and US Department of the Interior. Tested cleanup methods included burning some spilled oil. In that case, thick black clouds roiled off flames shooting up between masses of ice. In a second test, long "booms," essentially floating lassos, were towed behind boats to corral and concentrate floating oil. Then the oil could be burned or mopped up. Chemical dispersants were sprayed on oil. Skimmers—enormous yellow mops resembling the rollers with which you spread paint on a house wall—soaked up oil. Ground-penetrating radar detected oil films as thin as one centimeter, otherwise invisible within ice. Oil-sniffing dogs trotted over sea ice, their fur-flap-hatted handlers gripping their leashes as they sought hydrocarbons like bloodhounds following the trail of lost children. Tara the dachshund, wearing a green-and-yellow doggie vest, carried a GPS real-time tracking system around her neck.

Above the earth, meanwhile, SAR satellites searched in Arctic darkness and through clouds for escaped oil, able to concentrate on sea surface areas as small as a square meter.

And another kind of radar, suspended from helicopters, was flown at 20 miles an hour over spills.

The resulting report, "Oil in Ice," pored over by oil companies, governments and the Deepwater Horizon Commission, concluded that burning oil and chemical dispersants can be highly effective in ice if they are used quickly. That ice can retard the spread of oil on the surface. That darkness and remoteness can cripple an ill-prepared reaction system. That the corralling boom system can be effective in places where low ice conditions exist.

The report itself was the subject of heated argument in Alaska. Pete Slaiby quoted it often in speeches and hearings, using it to claim that oil-spill cleanup was highly possible in the north. Environmentalists countered that the tests had been done in too small an area, in good weather only and with prior knowledge of where the oil lay.

The Deepwater Horizon Commission came out somewhere in the middle. "The Svalbard tests were under extremely limited conditions. They raised some questions, answered some, and told us something about conditions," Fran Ulmer, commission member, told me. "Under certain conditions, ice can be helpful. Under others ice makes recovery impossible. If you are suggesting that the report shows that oil-spill response in Arctic conditions will be easy, forget it."

The commission also released a working paper on the challenges of oil-spill response in the Arctic and on Shell's plan in particular. It praised Shell for "going beyond" the standards for cleanup mandated by federal regulations. It concluded that cleanup in the Beaufort Sea would be "more straightforward" than in the Chukchi because the Beaufort "region is more developed and proximate infrastructure, so access to a spill might be easier. However," it said, "the Beaufort drilling sites are closer to both the sensitive shoreline and the areas traversed by bowhead whales and whale hunters."

In short, another argument.

Today in Dutch Harbor we would tour the *Nanuq*, the oil-cleanup ship shown in all those ads Shell had taken out in national newspapers, and the *Kulluk*, a relief drillship, to be dispatched to the Slope and standing by during drilling in case a blowout occurred while the company's *Discoverer* sank an exploratory well.

Plunging earthward we'd glimpsed both vessels—the blue-and-white *Nanuq* in the harbor, not far from the odd looking circular *Kulluk*, its derrick thrusting upward in the middle. The floating rig had been towed in from the Canadian Arctic, where it had drilled other exploratory wells for Shell. It was an older craft built in 1983, but the *Nanuq* had been constructed recently, specifically for work in heavy Arctic seas.

Pete Slaiby hoped that Shell's guests would "Ask questions. Kick the tires."

The rest of the view on the way down had been raw—thrusting peaks, sheer escarpments, the Makushin Volcano, which still smoked occasionally. Nature dwarfed any man-made structures but the structures were solid—a cozy-looking town and a cupola of a Russian Orthodox church; docks, piers and warehouses along shore; cranes, bars, paved streets, a Walmart-sized one stop superstore, and a World War II–in–the–Aleutians museum (Japan attacked the town in 1942). There was also a new community center with a heated Olympic-sized swimming pool, an immense dormitory for food-processing workers and a comfortable hotel—all in all a hardy foothold on the fringe of the Arctic.

The flight attendant opened the door and wind howled in as we filed outside to walk to the small terminal. One out of five flights to Dutch Harbor are canceled due to bad weather. The winter trip had been planned for weeks as part of Shell's long-range strategy to garner Eskimo support or at least blunt opposition. But now new urgency marked the day. With the drill suspension over at the Department of the Interior, and Shell's 2011 permit application alive again, it was now "crucial," Curtis Smith said, to keep Native Americans from joining national environmental groups in court. Better yet, to encourage them to speak up for the plan.

"There was a window of opportunity," Curtis said. "And we knew if we lost the stakeholders, we'd never get over the hurdles in DC."

Over the coming weeks BOEMRE would take public comments as it analyzed Shell's revised plan, in at least the tenth round of DOI hearings designed to solicit reactions to Shell plans since 2005, Curtis recalled.

"We used to live or die by these periods," Curtis said. "But then there's another and another. They never stop."

Between 2005 and 2010 just the EPA had held 200 days of public comment periods for the clean-air permit alone, Susan Childs told me in Anchorage in disgust. To this she added that the Army Corps of Engineers also regularly held public comment periods, as did NOAA and the Fish and Wildlife Service.

All present today were growing somewhat sick of public comment periods, the grinding, repetitive confrontations marking every step of federal processes. They did not mind the concept, just the endless rounds.

But you had to take each one seriously because even if it didn't help it could hurt. The process was so arduous that in some villages fewer people were showing up for meetings and even Mayor Itta's staffers felt twinges of sympathy for Slaiby sometimes. "He's in the permit gulag," Andy Mack, Itta's principal political adviser, joked. "You go in but never get out."

On December 9 Slaiby was back from the Hague, where high-level talks at headquarters had focused on various Arctic projects, not just Alaska. The other projects were moving forward. The company was bullish on the High North. In the coming years, Shell would be "working on our leases in Greenland," Pete said. "We'll be active in Russia, at Shtokman. We've got Eastern Siberia onshore. In Norway we're a player in different areas."

After almost three decades of business travel he was able to grab sleep on planes and had rested on the way back from Europe. He wore a red fleece jacket and carried a small overnight bag. The Shell people tended to stay together this morning, as did their guests. The overall mood reflected a comment Mayor Itta had made of Slaiby once: "I'll work with Pete, but I'm not about to go down to the beach with him, arms around each other, singing 'Kumbaya.'"

Itta had not come, as he was occupied in Anchorage trying to convince the Alaska Eskimo Whaling Commission to stay out of court. But Mayor Martha Whiting of the Northwest Arctic Borough, Itta's neighbor to the west, was present. She was a slim, dark-haired woman wearing designer glasses.

"I came to see these ships for real, not just in PowerPoint presentations," she said. She opposed offshore oil in principle but wrestled over whether it was incvitable. The Northwest Arctic Borough Assembly was about to change the wording of its resolution regarding offshore drilling from uniform opposition to something milder, but she was unsure what.

Guests ranged in age from their 30s to 60s and were equally divided by gender. John Hopson Jr. was a jovial, sharp-tongued whaling captain and director of operations for the Olgoonik

Corporation in the city of Wainwright, which sat on the Chukchi coast near Shell's leases and would be a staging area for offshore work if Shell's plans went through.

Hopson supported Shell generally—and the jobs the project would bring—but he had brought a tape recorder to bring back the words he heard on the trip so anyone in Wainwright could hear them.

Patrick Savok was tall, 33, a member of the Northwest Arctic Borough Assembly. He wore tattooed Chinese characters on his neck, and his stylish glasses and goatee made him look as if he would be just as much at home in an East Village, New York, nightclub. He carried a book, *Jesus and the Eskimos*, written by his grandfather's brother—the story of his family's heritage and relationship with Christianity. His home village was Kotzebue, also on the Chukchi Sea.

Dora Leavitt from Nuiqsut, on the Beaufort Sea, said her husband and two sons were whalers. She feared that if bowheads were diverted from migration routes it would endanger her men during fall hunts. That was because during those hunts, men go out in metal boats with outboards, not skin boats propelled by paddle, as in spring. Little if any ice covers water in autumn so there is no platform for camps. Hunters must venture out farther to reach whales. The farther they go, the more dangerous the trip. Boats can be swamped when storms come up.

The day had begun at 5:30 a.m. at Anchorage's Cook Hotel, where Shell had put up their Eskimo guests, when a tourist bus pulled up in lightly falling snow to take us to the airport. It was 11 degrees and streets were covered with packed snow. The mood was jolly at first. Old friends from different villages said hello.

"Hey, this is like a high school road trip!"

But the mood sobered during the flight despite a tasty bagel-and-fruit breakfast, when people began handing around the day's *Wall Street Journal*. A front-page article focused on the rising numbers of recent deepwater-oil-rig accidents or near misses that had occurred around the world.

Despite industry assertions that the *Deepwater Horizon* accident was a fluke, the paper said, statistics showed otherwise.

In 2009, the article said, in the US Gulf of Mexico, there had been 28 incidents where significant releases of natural gas or oil had occurred, or in which workers had lost control of a well. In Australia, just during the first half of 2010, 23 oil spills occurred. Britain's North Sea incidents were up 39 percent in 2010. The article reinforced the fear even among those who thought offshore development inevitable—and knew that Shell planned wells in shallower waters—that someday, a dreaded accident would occur.

The city of Unalaska spreads itself out along low areas between mountains. Two-lane shore roads curve past huge ravens and bald eagles perched on light poles, stores, garbage bins and shipping crates—three in a row atop a 7-Eleven, where the seedy setting made them look more like immense white-capped pigeons, urban creatures, than proud national symbols. Sort of like George Washington strolling on the 21st century Bowery.

At the Grand Aleutian Hotel we dropped our bags in rooms offering magnificent harbor views, and headed to the second-floor Makushin banquet room for a steam-table lunch and briefings by Shell staffers, most of whom had arrived on earlier flights. There were oil-cleanup men from Alaska Clean Seas and ASRC. Geoff Merrill would describe the state-of-the-art cleanup equipment he'd been buying. Shell's Micky Becker, a well known Anchorage PR consultant, had drawn up a strict timetable for briefings, safety talk, ship visits, dinner while watching company videos and finally questions and answers to wrap up the day.

But the schedule fell apart within minutes because of a rescue at sea and because Shell's guests had no intention of following company plans.

The *Golden Seas* was the ship that needed rescue. It was being pounded by 30-foot waves and 45-knot winds in the Bering Sea when the emergency call had come on December 3. The 738-foot-long dry-bulk carrier, heading from Vancouver to the United Arab Emirates, had lost its main engine 400 miles from Dutch Harbor in the Bering Sea. The engineer lacked parts to repair it. With speed reduced to two knots, there was no way to stop the carrier from being driven into rocky Atka Island unless another ship could tow it away.

Not only were the lives of twenty crew at stake, but if the ship hit land it would spill—in addition to its cargo of canola seeds—450,000 gallons of fuel oil, 11,700 gallons of diesel fuel and 10,000 gallons of lube oil. This volume of oil was greater than what had been spilled in most drilling accidents mentioned in the *Wall Street Journal*.

The carrier's peril mirrored predictions I'd heard from former Coast Guard commandant Thad Allen on the *Healy*, that as ship traffic in the north increases, so will the danger of accidents.

The Coast Guard had no ships or bases nearby and could not help the *Golden Seas*.

But Pete Slaiby could help.

That was because the nearest ship powerful enough to tow the *Golden Seas* to safety—if it could reach it in time—was Shell's Dutch Harbor–based *Tor Viking II*, an 18,300-horsepower tug/supply vessel leased from a Swedish company, Rederi AB TransAtlantic.

The *Tor Viking* had been scheduled to bring Shell's guests to the *Kulluk*, but Slaiby ordered the tug to head for the carrier instead on December 3. It was a risky decision, putting the ship in danger too, but after a 40-hour transit through 30-foot seas, the sleepless crew had managed to hook the *Golden Seas* to a towline and brought it back to Dutch Harbor.

Now the *Golden Seas* was saved, but the *Tor Viking*—still

maneuvering—was not available to take us to the *Kulluk*. Slaiby was proud of the rescue and used it to suggest that company ships could regularly assist in emergencies, if they were based in the north.

"If we have assets, they are always available."

It would not be a bad idea to have private vessels—especially ones designed for Arctic work—close by to help out in the opening Arctic, he said.

Slaiby told the group, "People charge us with being unable to work in the Arctic. We *are* able to work here."

No audience on earth could appreciate what Arctic sea rescue meant more than this one, but that did not alleviate the fears of drilling. As the initial briefings began, a barrage of unscheduled questions dealt Micky Becker's careful schedule another blow.

Mayor Martha Whiting asked, "The emergency helicopters. Are they strictly for emergencies? They're not going to be cruising around, are they? Because we have to be very quiet when we're beluga hunting."

Pete answered, "We will stay clear of where people are hunting. When we have to come onshore we will have subsistence advisers [Eskimos] on hand."

The mayor asked, "Will the names of those advisers be on your website?"

Whiting wanted to know also how many vessels in total would be coming?

"Ten to eighteen."

Questions came faster. The schedule called for a *Nanuq* tour at this point, but no one was budging from the banquet room. Slaiby and staff cordially answered all comers.

"How far from us will they be drilling?"

"How many choppers will be transporting people and how many times a day?"

"Where will search-and-rescue helicopters be located?"

On screen appeared flow charts of "risk mitigation" and "redundant blowout preventer control." Shell drilling superintendant Jim Miller showed a slide of the "Arctic response toolbox" and of the containment dome. The Shell people were experts on machinery. The audience was comprised of technical experts on ice.

"I hope some of *us* will get some of those 35,000 jobs," Mayor Whiting said.

Slaiby replied, "Without opening the door those 35,000 jobs will never happen."

"Other people get rich and we don't," Whiting said.

Finally it was clear with a return trip to Anchorage scheduled tomorrow that the questions had to pause if the group wanted to see the *Nanuq* in daylight. After a safety briefing we drove to the harbor and donned hard hats and started up a steel ramp to the *Nanuq*.

Behind me, Mayor Whiting said, "We've heard all this before."

What they had *not* done was tour Shell's vessels. For some, feeling solid deck beneath our feet gave drill plans a reality they had lacked before, they said.

One key to overcoming trepidation, Pete hoped, was plain talk. He knew that much of Shell's difficulty in returning to Alaska had initially been due to a misunderstanding of Eskimo power structure and ignoring of local concerns. It had taken years before Houston accepted that decision making in Eskimo villages was by consensus as much as possible. A smart executive did not talk only to a mayor or tribal leader but with an assortment of native corporations and whaler groups, any one of which, if angered, could go to court.

Now that error had been corrected, but this had not eliminated misunderstandings. It was still easy for company and North Slope people to interpret the same event in different ways. In the buildup to the trip, for instance, Shell staffers grew disappointed when Mayor Itta designated a replacement to come. That suggested in Anchorage that he regarded the trip as unimportant. The truth was that he thought it very important but did not trust his technological expertise when it came to analyzing what he would see. He preferred to send a substitute who knew more.

But when Shell turned down the substitute it fed suspicions among Itta's staffers that the oil people had something to hide, the irony being that the designee had just reported to Itta that Shell's discharge plans in 2011 looked fine.

Another potential pitfall was how the two groups regarded the technology they were looking at.

Richard Sears—adviser to the Deepwater Horizon Commission and a former Shell executive—once told me, "In Shell's culture, technology is who you are, how you get things done. When Shell has a PR problem often the root is that they try to explain technology associated with it and the audience doesn't care about that. They care about something else…Shell has a bad habit of misreading the audience and thinking, *Look, you're smart, I'm smart, let me explain this to you and then you'll think the way I do*. But not everybody values technology the way Shell does. And the Shell people go away thinking, *Why don't they believe me? I've been doing this all my life!*"

This sounded a lot like the way many North Slopers described their first meeting with Pete Slaiby after he arrived in Alaska.

On December 9 we heard a lot about technology while touring the *Nanuq*.

The ship was 301 feet long and had been built in 2007 for $100 million. The high-tech bridge included wraparound windows, contoured captains' chairs and computer displays tracking power from engines to propeller shafts and generators, electrical system, fuel.

Four enormous reels in back each carried 750 feet of inflatable booms for corralling oil on the sea. There were eight 40-foot-long rope mops to soak up spilled oil.

On the bridge, through which three groups of guests had been scheduled to rotate, but none would leave, Capt. Michael Terminal of shipbuilder Edison Chouest cheerfully answered more questions as Shell's Micky Becker whispered to me, nodding at the mass of casually intransigent visitors, "This is not what I envisioned."

"The *Nanuq* can hold position in 70 knots of wind and in a thirteen-meter sea. When I say 'hold position,' that's hands off," Terminal said. "This vessel can sit on position in that much wind and not move within one meter."

Outside, in December, snow drove in sheets across the harbor, stopped, started, stopped.

"What if the equipment fails?" asked Delbert Rexford, land chief from Barrow's Ukpeaġvik Iñupiat Corporation, who had led the saying of grace before lunch.

"Redundancy systems," came the answer. "Two generators and an emergency generator. If everything else goes we have that."

Mayor Whiting asked about sea trials that the ship had gone through. "Was the ship checked in the Chukchi or Beaufort in actual ice conditions?"

"Yes. In the Beaufort and Chukchi."

"What kind of ice thickness?"

"First year."

The close-up view nudged some visitors toward more acceptance of Shell's claims, but there was also a palpable sense of pressure building among those who felt helpless against their certainty that no matter what they decided, the ship would eventually be deployed.

It all came to a head that evening when Pete Slaiby showed the slightest frustration. With the briefings over, discussion turned to jobs and Pete rose at his table and suggested with mild reproof—as the audience sat at round banquet tables—that those who wanted to reap the benefits of drilling might help achieve them faster if they supported Shell, or at least stopped talking down company plans.

The words "talking down" lit the fire.

Mayor Whiting stood and her voice grew strident. *"Talking down?* Putting up a wall? It's about our life! *We're* gonna take all the risks. *We're* gonna get minimal benefits. Jobs will be there but not necessarily for us in a high-unemployment area. *You better believe we're going to be careful. We've been burned too many times."*

Her voice cracked. "We're stuck here. My people. Just because we oppose things doesn't mean we don't want you doing business in our region. We want to make sure with every ounce of power that we have, that even though we have no jurisdiction that our voices are heard."

Technicalities were forgotten and so, for a while, was any inter-play between groups. The Shell people became bystanders. The Eskimo group turned inward. Dora Leavitt spoke passionately of villagers who had gotten jobs at Prudhoe Bay and came home with terrible stories of being called "dirty Eskimo!" by white workers.

"That was talking down!" she said.

John Hopson Jr. chided North Slopers who refused to leave their villages for jobs in other places.

Patrick Savok stood and announced, "We need to go back to our communities and tell people to finish high school. There will be opportunities. We as parents and grandparents need to empha-size this. It is our responsibility."

Delbert Rexford relished the talk. "This is only the beginning," he said.

Shell's Geoff Merrill, off to the side, whispered, eyes rolling, "This is what happens every time."

———

The impossibly tall blond man who sat down beside me amid the commotion introduced himself as Ake Rohlen, director of Arctic services for TransAtlantic, the company owning the *Tor Viking*, savior of the *Golden Seas*. He'd flown in from Sweden to congratulate his crew and accept thanks from Pete and Unalaska's mayor.

The group gave him a long round of applause.

During a break Ake spoke to me in happy terms of the company's growing Arctic business. It was taking place partially due to gaps in US planning. He said TransAtlantic operated four icebreakers and was building more. TransAtlantic's icebreaker *Odin* was on lease to the US National Science Foundation in Antarctica. Since the United States had no functioning icebreakers except for the *Healy*, if the nation wanted to get supplies and scientists to polar bases it had to rent ships from other nations to do it. Ake was happy to provide this service at a good profit.

"The Arctic is not happening in the future. It is now," Ake said. "Look at icebreaking in Russia and in the Baltic as performed by Sweden and Finland. The state operates an icebreaker service. It's a tool for trade. But US icebreaking hasn't got a commercial enabling procedure."

Ake shrugged. "You have to think of yourself as an Arctic nation to institute these things. But I don't think America considers itself an Arctic state yet."

That was fine with him.

Before dawn Pete Slaiby got into a pickup truck with a driver and bounced off from the hotel for a close-up look at the *Kulluk*. It was anchored just offshore but there was no time to board it. In the dark, hard hat on, hands on hips, Pete surveyed the floating rig, clearly a man who would rather be drilling instead of making sure—as he was that week—that the EPA got another 700 pages of technical documents describing clean-air efforts. Or that his staff kept up with Department of the Interior drill-permit-requirement revisions. Or that he better remember to keep reproving tones from his voice during the rest of the trip.

The *Kulluk* seemed larger up close. Onshore in the dark pile drivers hammered as Pete gazed at the rig with the sort of appreciation with which a NASA engineer might regard an old yet well-functioning space shuttle.

On the way back, the driver, who was based in Unalaska, told Pete that occasionally Greenpeace activists show up in town to speak up against Shell. Greenpeace was greatly disliked in Dutch Harbor, he said. He glanced at Pete as if judging the reaction. It was unclear if he was merely conveying information or had a subtext. He repeated the message. "Greenpeace types don't do too well here."

Slaiby turned to the man. "We gotta make sure nothing happens to them. If someone gets overzealous and knocks someone else's teeth out, that's not good for us."

The driver did not mention Greenpeace again.

Unemployment across the United States reached 9.4 percent by that December. Millions of Americans had lost their jobs in the recession. To these financial drains were added the cost of wars in Iraq and Afghanistan. More than 22 percent of college graduates that year did not find employment. The national debt approached an astonishing $13.9 trillion. Federal revenues had plunged.

With Congress deadlocked over what programs to cut to find savings, the Treasury needed money, and not the kind you simply print more of but the sort that comes from having something to produce, including oil. Offshore production had been an issue in the 2008 presidential election. It was clear that it would be a bigger issue in 2012.

How many precautions would be necessary when it came to

allowing the drilling of one exploratory well in the Arctic, I asked Pete Slaiby as we waited to leave Dutch Harbor at the airport.

"I wish I knew," he said.

Also, was it really possible that this out-of-the-way harbor could one day transform itself into a Singapore of the north, or was it just a dream? I'd asked that question to Capt. William Rall of the *Healy* back in September, and he'd said it made sense to him. I'd asked local harbormaster John Days if Dutch Harbor could handle that much traffic.

"No problem."

Looking down as we took off—eyeing the snowcapped crests—I recalled hiking up the hills in July when they were green and rain-swept and looking down at the harbor. What had struck me was the resemblance to the lands around Panama City, the Pacific Ocean entrance to the Panama Canal. It wasn't tropical here but the geography was similar. I'd visited the canal earlier that year with the Coast Guard and stood dazzled on the deck of a cutter, looking out at the Hong Kong of Central America. The protected anchorage. The volcanic shore. The over 200 high-rise buildings under construction. There was no snow, of course, but the Pacific Rim topography was not dissimilar to Dutch Harbor's, and rising over the island on December 10—heading back to Anchorage—it was not a stretch to mentally superimpose more warehouses and more ships onto the protected island and harbor below.

After all, by December of that year, Alaska's Republican senator, Lisa Murkowski, had introduced a bill in the US Senate asking for funding to study locations for a US Arctic port.

Maybe it will be Dutch Harbor.

Maybe Barrow, Scott Borgerson had mused.

Mayor Itta was in a jolly mood as he waited for Pete Slaiby in Anchorage.

At 4 that afternoon Itta sat in a North Slope Borough office conference room on C Street, relaxed and voluble in a blue dress shirt open at the collar, sleeves rolled up. He was feeling confident for a change after what he judged a successful week in town. He had come away from a meeting with the Alaska Eskimo Whaling Commission believing that the whalers would from now on stay out of legal action to stop Shell's 2011 plan. If the clean-air-permit appeal went through, the whalers—also believing that some off-shore drilling was inevitable and also somewhat mollified by changes in Shell's plans—would go along with it.

He had also met with angry Gov. Sean Parnell and assured him that he was not actively opposing Shell. Itta as usual was in the middle, but things seemed more workable that day.

Pete Slaiby seemed uncharacteristically glum walking in. The cumulative weight of recent weeks was getting to him. Things started cordially as he told Itta the story of the *Tor Viking* rescue and passed along words from Dutch Harbor's mayor Shirley Marquardt—which she'd used in a speech at the Grand Aleutian Hotel last night—about how the town had grown from a population of 400 to 4,000 and prospered after industry got a foothold there.

Edward had no problem with that. "We have more in common than differences," he told Pete. "I have to say we started off a little tense last spring. I have no desire to go through that again."

The men were getting to know each other better, not socially, but in the way of veteran campaigners who have something in common after a while, even if it is their differences of opinion. The mood between them had softened since the *Deepwater Horizon* had blown up that spring.

"We don't have a lot of comfort that we'll get to drill in 2011," Slaiby admitted. It was the first time I'd heard him say it.

Suddenly Itta was trying to comfort the oil man.

"I assured the governor that we are working together. It would behoove us," Itta said, laughed and added, "I love that word 'behoove'... to tell the governor that we're together more than on the federal side."

Back in May Itta had blown up when the oil man suggested that he explain Shell's position to his people. Now Itta was asking Slaiby to assure the governor that the men were not so far apart.

SLAIBY: We're in the middle of commercial negotiations. BP's performance in the Gulf is going to cost us dearly... I ante up and ante up and I never get a handout... If we're not going to drill in 2011 I don't want to spend more money.

ITTA: I can imagine the position you're in, people breathing down your neck. But Pete, I want to keep moving forward. I'm happy we've gotten to a good place on the science agreement. I'm with you on respect to your ass. I understand where you're coming from. I'm not going to sit here and say I empathize 100 percent, but I recognize you have constituents who are in different places and we will work with you to take a little bit of burden off your shoulders. The system here is absolutely broken with respect to the lack of certainty over what is going on.

SLAIBY: There has been a whole range of politicals making decisions without understanding the permitting process or trusting results coming out of their own offices... These guys are fly-specking documents. There's going to be some vulnerability in any 300-page environmental assessment or 800-page environmental impact statement.

ITTA: Imagine if I had been of the other stripe. If I had opposed the shit out of all of this until I'm a dead man. That's the other option. You should be thankful that I've convinced our whalers to go along with my approach.

The conference room had no windows so scenery consisted of blow-up black-and-white photos of Iñupiat elders, and hunters saving a whale trapped in ice by cutting out a rectangular air hole through which it could breathe. This was the rescue that had inspired the film *Big Miracle* that had brought actor Ted Danson to Alaska, where he'd attacked Shell.

Slaiby shook hands and left.

Minutes later Karin Berentsen of Statoil walked in. She was the Norwegian state oil company's "Alaska stakeholder manager," a term meaning someone who deals with the locals. The friendly and slightly reserved blonde had recently moved to Anchorage, wore pinkish glasses and a yellow scarf, and was meeting Itta for the first time. She'd brought along Bill Schoellhorn from Statoil's Houston office. He was the company geoscience manager for global exploration in North America.

As they sat down across from Itta and aide Andy Mack, the sense filled the room of a parade of oil companies waiting in the wings. ConocoPhillips. Total, from France. Repsol, from Spain. Eni, from Italy. And any company that coveted one of those little square boxes that divided up the seas off the North Slope on the map in Itta's conference room back in Barrow.

Karin Berentsen handed over another carefully prepared company handout showing future plans to drill offshore. A snowy owl peered out from the cover. NORWAY—ALASKA 60 DEGREES LATITUDE NORTH, read a headline. It implied that both locations shared a latitude and therefore mutual interests. Once again I was looking at the North Slope divided into small boxes. Once again, in "Statoil's seismic survey area," the boxes extended out into the Arctic seas. Once again I saw photos of a blue-and-white oil company ship, and this one, the *Geo Celtic*, was attended by "environmental support vessels." The "Petroleum Industry in Norway" sheet showed waters off that nation divided into boxes too, and a mass of thin red lines—functioning gas pipelines—running from sea to shore.

Statoil owned leases in the Chukchi and Beaufort Seas but had not yet made some of the concessions that Shell had when it came to environmental precautions. This was one reason why Itta was being insistent with federal officials that Shell's promises and private agreements be incorporated into all lease requirements. If predictions were right about the bounty off the North Slope, someday there might be hundreds of wells offshore—producing wells and not exploratory wells; full-time wells, not seasonal ones—and to Itta's mind the safety and environmental rules better go into effect now for all.

The meeting began as cordially as a diplomatic cocktail party. Bill Schoellhorn assured Itta that, "I will commit some time up here. We'll always be open with you." Itta responding, "I commend your approach and welcome you. The North Slope needs to be part of it from the beginning." Blah, blah, blah. There were smiles all around as if this was going to be easy in the end. Bill saying, "We're starting to look at specific areas. I think 2014 is the target date for us to begin our drilling."

Itta took a swipe at the US government. "If MMS had been doing its job from the beginning, half the issues we have today, we wouldn't have," he said.

The meeting remained smooth until Itta left for another appointment and the North Slope end was taken up by Andy Mack. As long as the discussion stayed on generalities, everyone kept smiling. But Mack got down to specifics. He wanted assurances that if Statoil found oil, the company would commit to transporting it by pipeline, not by ship. The North Slope could tax a pipeline on land but it could not tax a ship.

Suddenly the smiles grew tighter on the Statoil end.

"A good idea but I can't exactly say," Karin said.

When Andy pressed she demurred again. "Oh, we have to find the oil first."

Once again, a North Slope/oil company dance of interests had begun.

After the Statoil reps left, Andy pushed his chair back, remembering dozens of meetings over the years with oil company representatives. "Time after time I see new ones come in, give a presentation and leave, thinking, *That went well.*" He added with an impish grin, envisioning the fights to come, "Then I find out what the North Slope people *really* thought."

Shell's Christmas party started at 5 p.m. at the Cook Hotel on December 13 as a dusting of fresh snow powdered what already lay atop the sparkling city.

Guests included newly sworn-in Lt. Gov. Mead Treadwell; Geoffrey Haddad, vice president of exploration and land for Conoco-Phillips; Capt. Mike Terminal from Edison Chouest; and Shell vendors and contractors. No one from the North Slope was there.

Downstairs, arriving guests paused in the lobby to admire the hotel's annual holiday season gingerbread city, a large tabletop display produced by pastry chef Joe Hickel, son of former Alaska governor Walter Hickel, also a former secretary of the interior. Gingerbread Anchorage spread out in sugar and chocolate, a toy-sized metropolis where visitors could literally sniff the homes of chief citizens, labeled in milk chocolate. No home in the display belonged to an oil man. Outside, expensive cars regularly pulled into the semicircular driveway and owners got out with digital cameras to take shots of miniature versions of their homes.

Upstairs, revelers dined on roast beef, iced shrimp, steamed vegetables, crab cakes and fresh rolls. There were mint petit fours and wine, beer or soft drinks at a bar.

Many present looked forward to vacation trips during the Christmas holidays. Pete Slaiby would head off to Brazil and a visit to Rejani's family.

"I can use it."

Mayor Itta would spend Christmas on the North Slope.

Micky Becker, Shell's events arranger, clinked a fork against a wineglass to signal for quiet as Pete stepped up to a podium to convey holiday greetings. His exhaustion was gone. He was a welcoming host. In his brief remarks he made reference to prior holiday parties where he'd also appeared and hoped for a productive year in Alaska for all, but then he had not had one.

"Sometimes I feel like I'm in the movie *Groundhog Day*," he joked, referring to the film where a TV weatherman is doomed to repeat the same day, the same efforts, over and over to no avail. The line drew laughs but also a sympathetic quiet.

By the bar I met Geoffrey Haddad from ConocoPhillips, who told me he didn't think Shell had a chance of drilling in 2011. It simply, he predicted, would not happen.

"Shell leads with their chin," he said.

Pete was on vacation on December 30 when the EPA appeals board finally handed down its decision. It wasn't good.

He didn't get the news until he was back from Brazil. The board was sending the clean-air permit application back to the drawing board, *remanding it*. The three-person panel, based in Washington, DC—*the EPA's own board*—ruled that the agency had, after years of analysis, failed to consider "environmental justice" impacts of air pollution when it approved air permits for Shell's offshore exploration in the Chukchi and Beaufort Seas.

The ruling charged the EPA with erring primarily in two places, first when deciding at what point Shell's drillship was to be considered a source of air pollutants. The agency considered the ship a source after it dropped anchors to drill. The agency's remand board ruled that it should have been considered a source earlier.

Normally, ships moving at sea are not covered by the Clean Air Act.

The board also ruled that drilling might cause disproportionately high and adverse health effects among natives, so it wanted an expanded analysis. When the permits were issued, agency staffers had accurately applied current standards but not a new EPA rule due to go into effect two weeks later, mandating a maximum one-hour limit for emissions of nitrogen dioxide, a pollutant produced by internal combustion engines and thermal power stations. Now the remand board wanted the EPA to reanalyze the whole application in light of the rule or explain why it should not apply.

After all, although the one-hour emissions standard rule had been described by EPA administrator Lisa Jackson in a release as designed to "prevent short-term exposures in high-risk NO_2 zones like urban communities and areas near roadways," and although the village closest to Shell's drillship would be 70 miles away, whalers had complained that their boats would be closer.

The board declined to rule on other complaints, saving them for later, which suggested that if the first concerns were mitigated, more rulings would be required one day.

Geoff Merrill, the old Coast Guard sailor, summed up his feelings with a seaman's eloquence. "We've been fuckin' yo-yoed around."

Pete Slaiby was furious. Walking into the office on January 2, first workday of the new year, he was white, jaw clenched, but the first thing he did was to reassure staffers that they had done a good job, that their jobs were not in jeopardy and that Dave Lawrence in Houston was not angry with them.

He was not ready to give up.

"Maybe it can be fixed," Curtis said.

Maybe there was a way to get the remand board the information it needed, solve the problems and in a timely way keep 2011 alive.

"You know, it's like after a car accident. The first question is, 'How bad is it?'" Curtis said.

Aggravating emotions at Shell was the fact that *Congress* had not established the EPA review board. It did not exist by law. It had been set up inside the agency.

"The EPA is supposed to use rules that legislators put in front of them, not generate their own," Slaiby said.

Slaiby got on the phone with EPA staff in Seattle. These were the people who had done the analysis that had just been thrown back into their faces by their own board. Time after time over the last few years Shell staff had offered to work with the EPA in drawing up permit applications, but the offers had been rejected because "they didn't want to be too close to us," Susan Childs said, which was fine with Slaiby *as long as EPA staffers did their jobs right*, which, according to their own internal judging system, they had not.

On that call Susan Childs noted how Pete's voice kept getting lower, how his fingers started shaking. He was jabbing them at the speakerphone as bureaucratic soothing tones emanated from it, ratcheting his fury up further, the people on the other end telling him to be calm and take things slowly. Telling him that it was doubtful that work that needed to be done could be completed in time for Shell to drill in 2011, which had now arrived.

Slaiby also got on the phone with Shell lawyers in Houston, and with Mark Stone, the lawyer in Anchorage, and the legal staff began analyzing the ruling, looking for a way to satisfy the new requirements fast.

Pete—at home—went sleepless at night, reading in the wee hours to distract himself.

He arranged another trip to Washington, to see Brian Malnak and talk to Alaska's senators again.

Back in the office, on his own, Curtis began working on one more press release announcing failure, hoping he would not have to release it.

Pete thinking doggedly, *Somehow there has to be a way.*

Chapter 11

Winter in Barrow, January 2011

Barrow's winter darkness lasts from mid-November until January. "Twenty-twenty" weather prevails during many days, subjecting the city to temperatures of twenty below zero combined with a twenty-knot wind. That lowers the chill factor to 70 below zero. Wind claws at your skin and knifes through parkas. People staying outside too long might show white patches on their cheeks—dead skin. Alcohol abuse goes up and hunting time down. Delbert Rexford went out after caribou by snowmobile, stayed out too long, and came back with frostbite in his lungs.

Barrow assumes an awesome beauty in winter. Ice sheaths the frame houses and the community center and stretches from shore all the way to the North Pole. Snow ripples across ice-covered Middle Lagoon, bunches on hummocks on the tundra, forms sastrugi—tiny parallel wavy ridges—and laps like sand around pilings supporting homes and offices, or overturned boats. It sparkles like diamond dust in the glow of streetlights. Living and inanimate things wait for the sun.

Natural light finally peeks through in January as a low, luminous band of pink spreading across the horizon at around one in the afternoon; an hour-long taste of color creeping into the world. Day by day the light stays longer. On weekends Geoff Carroll likes to take out the dogs. He's the last musher in town. His big Greenland huskies live penned in a chain-link warren of doghouses behind his two-story home on Yugit Street. They howled and leaped excitedly as we unchained and harnessed them to the wooden sled for their Saturday run. There was no steering mechanism to control the sled, just an iron footrest in back, a brake. The musher— standing on the runners—could step on the brake, driving it into the snow to slow the dogs.

Geoff, 60, walked with the rolling limp of a man who'd recently undergone a hip replacement. He had the ruddy face of a 49er, a benevolent manner and easygoing attitude toward outdoor travel. Sometimes he ran the dogs to outer villages while working as a wildlife biologist for Alaska's Department of Fish and Game, where he monitored North Slope land mammals: moose, caribou, wolves and musk ox.

He could fly on these trips but preferred the beauty and solitude of the wild. The dogs could take him up to 60 miles a day. He dug snow caves at night for sleeping. A trip to Wainwright took two days, he said.

Geoff steered by vocal command. His voice carried loudly in the frosty air. "Hut-hut-hut!"

The sled rammed forward at twelve miles an hour, across streets, a lake and onto tundra, heading toward the raised natural gas pipeline bringing energy to town. At least we thought we were headed there. A whiteout started. Civilization disappeared.

Eleven tails wagged like crazy as we floated through white.

"Want to try it?" he asked.

I stood in back and shouted, "Dee!"—turn right—and eleven dog heads whipped around as if the animals thought, *Who the hell is that guy? No way I'm listening to him!*

I tried "Haw!" Turn left!

The sled continued straight at six miles an hour.

Like most people in town, Geoff waited to see if Shell's Alaska Venture in January would find a way around the EPA remand. He was not keen on seeing drillships offshore and feared a spill. He stressed that this was a personal opinion. He did not speak for his department.

Geoff took over and shouted "Dee!" and his nine-year-old lead dog, Xena, turned right. Through blowing white, the vague raised form of the pipeline appeared.

The High North had appealed to Geoff from the first moment he saw it. He originally came to Alaska from Wyoming with a brother and friend for "a quick summer trip in the 1970s, but I stepped across the border, looked around and thought, Man, this is where I want to be."

He ended up helping on the original bowhead whale census, which resulted in the International Whaling Commission reversing the hunting ban for natives. Later he worked as a borough biologist, during which time a polar bear almost killed Geoff out on the ice.

"It happened in camp during a whale census."

The bear had been attracted by smells from a cooking tent and the cook, sensing a presence, turned at the stove to see the animal coming through the flaps.

"He yelled at it. It didn't respond. He picked up the hot frying pan and threw it. It smacked the bear on the nose and it backed out," Geoff said.

Angry, the bear circled around and headed for two sleeping tents behind, one containing Geoff. Woken by the yelling, he was already half dressed. Census takers always slept that way in case the ice broke off beneath camp and they needed to get away fast. He walked out of the tent carrying a Remington shotgun, but it was not loaded. By the time he saw the bear it was too late.

"I tried to get a shell cranked in and the safety off and it jammed. The bear was coming. I used the shotgun as a baseball bat and shattered the stock on his head. I was surprised. It knocked him over."

But the bear stood up.

Geoff, helpless, was aware that someone else had come out of the adjacent tent and this second man "had his act together a little better with a shotgun."

When the bear charged, the man fired, killing it.

"I looked down to see two sets of tracks. Mine ended three feet from the charging bear's."

Geoff was the kind of man who quietly bit off outdoor challenges to see if he could beat them. He made international news in 1986, when he completed a 56-day, 500-mile-long trek to the North Pole from Canada's Northwest Territories, part of the first group to reach the Pole on foot without resupply since Robert Peary did it in 1909.

"Every night we'd sweat into the insulation in the sleeping bags and it would freeze. The bags would get heavier. After a while that gave you a bag full of ice weighing 30 pounds. You'd beat it with a stick to break up the ice. But the insulation worked even if wet. You'd crawl into the bag of ice cubes and warm up."

So when it came to personal challenges, he was deeply appreciative of the Arctic. But when it came to challenges from development he was of a different mind, glad that courts had slowed oil drilling even on land. In the National Petroleum Reserve, he said, due to an Audubon Society suit, the US Bureau of Land Management reanalyzed a plan to lease sensitive areas and "decided there wasn't as much oil as they'd figured, and dropped the plan."

"In this kind of work you can win fifteen times, but if you lose once, you lose everything," he said.

The whiteout worsened. Geoff pulled out a handheld GPS and realized that we'd been going the wrong way. The temperature plunged, but if you kept moving you stayed warm. Thanks to technology the city appeared and we bumped back past the airport, onto Geoff's road and down to his backyard, where, uncoupled, each dog received two foot-long frozen fish as a meal. Geoff catches them in nets, 200 at a time, he said.

By 2011 Geoff had been married into the North Slope for 24 years. His wife, Marie—like Mayor Itta—had attended the Bureau of Indian Affairs school at Mt. Edgecombe and was a former executive director of the Alaska Eskimo Whaling Commission. She also fretted about Shell's plans. Geoff's high school student son Gunnar had completed a science project in which he'd tried to burn off spilled oil in water, testing Shell's claims that this was an effective cleanup means.

Only 34 percent of the oil had been cleaned in the best of circumstances, a statistic found highly amusing by the BOEMRE official who told a Barrow audience that winter that the $20 high school science project had come up with the same results as the multimillion-dollar European cleanup study.

Geoff, in describing his courtship of Marie, said that she was impressed initially by the story of his "hitting the bear over the head, and when I went to the North Pole that kind of won her over."

I asked if he recommended thwarting polar bear attacks as a way to meet girls.

"Whatever it takes," he said.

Jason Herreman was interested in polar bears that month too, and any work on bears might affect Shell's plans. Jason's experiment required guns, barbed wire and tweezers.

The North Dakota–born biologist started up a Borough Wildlife Department snowmobile and we headed north from BASC toward the bowhead graveyard at the northernmost tip of North America, the headland where the Chukchi and Beaufort Seas meet.

Fit and blond, Jason wore wind overalls beneath his parka, soft warm boots, heavy mittens, a rabbit fur, snap-up Russian-style hat, and a Ruger .480 in a chest holster, just inside his parka, for easy access. It was loaded with regular bullets to be used as a last resort if we were attacked. Normally bear guards traveling with researchers carry a shotgun loaded with deterrence rounds; cracker shells and beanbags.

We bounced across tundra. Wind whipped up powder as the machine lurched over hummocks and crashed down inclines, the tread gripping snow or skittering over ice. Newcomers to the Wildlife Department occasionally find themselves beneath overturned snowmobiles. Dr. Murali Pai of India, the latest arrival, dislocated his shoulder beneath one in January.

Nine bone-rattling miles later I saw what looked from a hundred yards ahead to be a Stonehenge-style prehistoric monument—thin, long, curled shapes jutting fifteen feet up from the snow. Closer, I saw that the arced shapes thrust from a large pile that might have been crumpled boulders, and closer still the shapes assumed the geometry of curled finger bones, as if some giant had died and the flesh melted off his hand.

They turned out to be whale bones—bowhead remains, brought here by bulldozer—scattered and cracked open by hungry bears. The jutting shapes were ribs. The bone pile was here to keep bears away from town. It had not initially been intended as an experiment but to protect Barrow residents.

But the bone pile might help answer questions affecting national policy in the Arctic, including where pipelines might one day come ashore carrying oil.

That's because polar bear numbers are dropping around the world, scientists say, as sea ice that they use as hunting platforms melts away. Computer models predict further decline by combining melt rates with bear populations, and partially based on these predictions polar bears have been designated by the Fish and Wildlife Service as threatened in the wild.

If the counts are wrong, so are the predictions. If the predictions are wrong, so—say critics—are the laws.

On the North Slope the new rules would have immense day-to-day consequences. Labeling bear habitat as critical bars it from being developed. Immense stretches of shore go off limits to new roads, runways or oil pipelines. And limiting new pipelines could mean limiting potential tax income for the North Slope.

It was therefore crucial to get Mayor Itta proper bear counts, and that's where Jason's work came in.

Approaching the bone pile, we stepped over splashes of silvery bear excrement. A single strand of foot-high barbed wire circled the pile. As a barrier it was a joke, but Jason said the wire had a different purpose.

"The bears leave hair on it. That way we get DNA samples without having to tranquilize the animals."

A half mile west, two polar bears made small by distance eyed us. Heads down, they kept going.

Jason circled the "hair snare" wire, picking up hairs stuck there with tweezers. He deposited the hairs in a plastic vial and then repeated his route and tautened loose wire so it wouldn't collapse the next time a bear pushed through.

Identifying specific bears by DNA, he said, helps the department make more accurate population estimates.

"We have an argument between what our hunters see and what the Fish and Wildlife Service says. The hunters feel there are more bears out there than population estimates.

"There's a large percentage of bears that the Fish and Wildlife people and USGS have captured and put tags on. They've been sampled genetically. So if the USGS says we have 70 percent of the population marked, and *our* study finds 70 percent of the bears at the pile marked, we're probably in agreement on numbers. But if, say, we only find 30 percent correlation, maybe the hunters have a point."

Jason tended to think federal numbers accurate, especially in the better-studied southern Beaufort Sea, but he was open to any answer. He said he felt no pressure in town to come up with results favoring one view or the other.

"People here want to know the truth whatever it is."

The bottom line was that once again, changes in the Arctic were outstripping the speed with which science could track them. Governments and corporations needed to make decisions, but which ones were prudent and which were rash? How long could you delay decisions while waiting for studies? This was the dilemma faced in Washington because choosing to make no policy—with ice melting so fast—often represented policy making by default.

In January, some members of Itta's Wildlife Department still advised him to fight all offshore exploration until more science could be done. But when Itta asked if they had proof that sea mammals would be endangered by Shell's plan, the answer was no proof.

Itta stayed his middle course on oil exploration.

He headed for Washington in January to try to free parts of his coastal areas of the new threatened-bear-habitat rule.

Many people across the North Slope held their breath that month, waiting to see if 2011 would be the year that drillships finally appeared. But for others, short-term at least, daily life went on unchanged. High school students were preoccupied with an upcoming basketball tournament. Boys' and girls' teams would be flying in. The Borough Assembly voted money for a *kivgiq*—a winter dance celebration—and invitations went out to native groups around north Alaska.

On the police end, the usual winter upsurge of alcohol abuse was continuing. Barrow is "damp," meaning alcohol can be consumed but not sold. People who've been cleared by police (you can't buy alcohol if you have committed a crime) pay immense amounts to have bottles flown in and stored in a small frame house across from the airport. There they show ID and pick up the delivery. There are limits for how much you can order; 20 liters of wine a month, 4.25 liters of distilled spirits, 11.25 gallons of beer. Anchorage and Fairbanks liquor store 800 phone numbers are posted in the building. A permit to buy liquor costs $100, on top of which buyers pay a $25 fee for handling and a 3 percent alcohol tax and a $150 freight fee. This is still cheaper than the cost of bootlegged liquor in outer villages. There, police told me, a $10 bottle of Canadian R&R Whiskey in Anchorage can cost $300.

In the little house, city clerks dispensed alcohol. The back room was piled with boxes of Miller Draft beer, wines, spirits, Pete's Wicked Ale.

"There used to be a bar at Pepe's until the city went dry," said Arnold Patkotak, the clerk.

At Pepe's it was business as usual too that month.

Customers lingered over meals in the booths and three dining rooms, talking about families, gossip, arrests, hunting. If Barrow were a TV series, Pepe's would be the restaurant where residents habitually gather for breakfast. The place where everyone knows everybody else, and any issue worth hearing about gets aired.

A blizzard was coming in, but "We never shut down. Well, we did for one day in 1987 when the drifts were taller than me," said Fran Tate, the owner.

Pepe's sat catty-corner across from the ASRC building, the Wells

Fargo Bank and the police station, down the block from borough offices and a few windy steps from court. At 87, Fran still loved greeting up to 300 customers a day served by her seven Mexican cooks. They rotate in shifts.

"I never learned how to cook."

Fran was small and blond and had a girlish demeanor, open face, quick wit and iron will. She could veer in seconds from rampant generosity to disdain if a notion struck her as stupid. The festive ribbons in her hair befitted Pepe's—functionally trailer-like outside, but cheery and cozy inside, with bullfighter murals on walls, a fire pit in the main room, strung-up chili pepper lights in reds, yellows, and greens, and a sign by the cash register: THE MANAGEMENT HAS THE RIGHT TO REFUSE SERVICE TO ANYONE, ESPECIALLY CLARENCE WHEN HE IS DRUNK.

Years ago one Hollywood producer actually wrote a Pepe's-based sitcom, which started with a visitor walking in to see Fran and staff dancing in a conga line.

The show did not reach the screen.

"Shell people?" Fran said. "They stick out like a sore thumb. White people with a tie on. I say, 'Where you folks from?' When one says, 'Houston,' I say, 'Shell Oil?'"

Fans of Chicago's WGN radio knew Fran as the "Barrow Bureau" for the *Steve and Johnnie Show* in 2011. Her regular call-in stints began years earlier when the hosts decided to find the coldest place in the United States so they could poke fun of Chicagoans complaining about winter. Phoning the Barrow police, they were directed to the restaurant. Fran still took calls about once a month.

"They ask if we celebrate Easter up here, and I say," she said, screwing up her face, "yeah, it's part of the US, isn't it? They say, Do you have an Easter Bunny? I say, I *am* the Easter Bunny. At school I hand out candies. They say, Do you hide Easter eggs? How do you find them?" She rolled her eyes. "*'Cause we color 'em, just like you do!*"

"Oh, they ask all kinds of questions. They can't believe that you see polar bears across the street. A few years ago the ice went out and 42 bears were stuck here. They can't believe that whales weigh a ton a foot."

A graduate of the University of Washington in electrical engineering, Fran initially came to town to work for Husky Oil on the National Petroleum Reserve. She bunked at the old Navy base with 39 guys—"Not so bad." The bigger problem was toilets. "They were gas fired and disintegrated waste, but as soon as you finished you had to get up quick or you'd get burned. It wasn't leisurely.

"People asked me, 'How can you stand it?' Well, I was born in the Depression. My folks were from Europe. We lived in a tent and had an outhouse half a block away. Up here honey buckets in the house were an improvement. I came from a town where you could walk down the street and people would say, 'Hi, how are you?' You could answer, 'I'm gut shot!' And they would sing out 'Fine, thank you,' and walk past. Here you wait for an answer. Grandma is okay. The kids got pneumonia. You know everybody. It's real."

Deciding to stay, she drove a truck delivering water and owned a temp-secretary business before turning to food.

Eleven banks turned her down when she asked for loans to start a restaurant. Finally she overdrew her checking account by $11,000 to buy kitchen equipment.

Luckily, Pepe's caught on.

Fran hit the national media in the 1980s, in her case when a *Wall Street Journal* reporter stepped into Pepe's for dinner one night. He had not expected to find a Mexican restaurant in the Arctic. ALASKA'S FRAN TATE BECOMES A BIG WHEEL IN BARROW: ESKIMO TACOS ON TUNDRA, the ensuing front-page headline read. The piece quoted one of the cooks recalling his decision to accept a job offer in Barrow: "I thought the restaurant would be a big igloo," the man said.

The piece was so funny that NBC's *Tonight Show* staffers

phoned Fran, preinterviewed her and asked her to come on the show as a guest. They sent first-class airline tickets.

"I didn't like first class. My feet can't touch the floor from those seats."

Told to bring something to show host Johnny Carson, Fran selected a walrus penis—a long solid shaft—which he examined on camera. "What is this? A whale bone?" he asked.

"Oh, let's just say every male walrus has one."

Johnny put it down. "This makes me feel pretty insignificant," he quipped.

Initially scheduled for a six-minute segment, Fran heard the producers saying, "Go for it!" when time was up. She ended up on camera for 23 minutes and the next night, at the pricey hotel where she was staying, where the waiters had ignored her earlier, she found that they massed her table like Texaco servicemen in old gas station ads.

These days Pepe's catered public events in Barrow, including Eskimo Whaling Commission meetings over oil issues or Borough Assembly functions over oil issues. Pepe's sent over trays of hot lasagna, enchiladas, salads, ribs. With her oil-related profits Fran traveled to Chicago each Easter to donate toys at Children's Memorial Hospital. At home her generosity extended to new moms in town, who, bringing in their infants, got a baby outfit for free.

"Sometimes one of my sons says, 'There goes my inheritance,' and I tell him, 'You ain't gettin' nothin'.'"

Fran's politics seemed to favor people she liked and oppose anyone she deemed phony. One time she saw a recognizable face at a table that turned out to belong to a Wall Street financier accused of fiscal impropriety on the news. He was in town "trying to get his fingers in our pockets. *Wall Street!*"

"I was pouring him coffee and he said, 'I saw you on the Johnny Carson show!'"

" 'Yeah, I heard *your* name on TV too,'" she said. Smiling, Fran saw it hit home.

On local KBRW radio, the most popular call-in show never mentioned oil that month—or Washington, DC, or Arctic science, or the EPA or even whales. It was the Birthday Show. Since Barrow had a single station, anyone listening to any radio—Thai taxi drivers like Chairat Jarupakorn, 50; city employees in their snowplows; BASC mechanics and scientists in labs each night—heard the nonstop felicitations to children, parents, neighbors, cousins, friends. You didn't have to know the people personally to appreciate the goodwill. Richard Glenn's wife had first heard his name on the birthday program, on the day they met. It was impossible not to smile while listening.

"Birthday program! You're on the air!"

"Happy 14th birthday, Sheldon. Here's your mom!"

"Happy 19th, Kaitlyn, from your dad. I love you."

"A *very* special happy birthday to my grandson."

On and on. At the heritage center, in kitchens in outlying villages, over the hiss of Arctic wind the good wishes went on.

"Happy birthday from Uncle Max! *Hola!*"

I asked Mike Lane—ponytailed, bespectacled station manager who said he'd grown up as the "kind of kid back in Detroit who sat around reading the encyclopedia for fun"—if I could go on air and ask the live audience if anybody wanted to talk about Shell Oil. Mike laughed. He'd hosted the show for a while, he said, and tried to tell stories to the audience, chat, entertain, do the things that radio hosts always do.

"I got a flood of calls telling me to shut up. *We don't want to hear you!*"

I dropped the plan to seek opinions on the show.

"Happy birthday! Be nice to your mom! Love you!"

That January the Deepwater Horizon Commission's report came out in Washington. It blamed the oil industry and federal oil regulators equally for the explosion in the Gulf of Mexico. It called for an overhaul of the whole system, and for more science and careful planning in the Arctic.

The report called BOEMRE "underfunded" and filled with personnel who are "often badly trained."

It called on the nations of the Arctic to establish strong international standards related to Arctic oil and gas activities that would require coordination of policies and recourses. Considering that the US had not even been able to ratify the Law of the Sea Treaty, this recommendation had little chance of being instituted.

The report advocated the creation of an industry-run organization to establish better drilling practices.

It called for more funding for key regulatory agencies that oversee oil-spill response and planning, including Interior, Coast Guard and NOAA.

It recommended that the Department of the Interior should include NOAA in decision making about where and how to conduct offshore lease sales.

"Science has not been given a sufficient seat at the table," said commission co-chair Bob Graham, a former senator from Florida. "Actually, that is a significant understatement. It has been virtually shut out."

More regulation, it seemed, was being called for.

A spokesman for Interior responded that BOEMRE had already made important changes, creating a special investigations unit to look at industry wrongdoing, and separating the leasing revenue part of the agency from safety and environmental enforcement.

But more was needed, commission co-chair Bill Reilly told an audience in Washington's Reagan Center that January. "The future of oil exploration is in deepwater and it may also be in the shallow waters of the Arctic." The Arctic was, he said, "a punishing environment with different challenges, precarious wildlife species, and a little-understood ecology..."

Reilly praised "several companies with exemplary safety and environmental records...I would challenge them to use their example and influence to champion a strong, industry-wide culture of safety."

He did not mention the names of the companies.

Later he told me the list included Shell.

Edward Itta and Richard Glenn were headed to Washington in January for meetings with agencies, a talk with Senator Begich and an appearance with Pete Slaiby on a panel at the Reagan Center, at a conference on oceans policy. They would discuss balancing protection and development in the Arctic. Richard brought maps along showing newly designated polar bear critical habitat on the North Slope superimposed over the East Coast of the US.

His plan was to use the maps to dramatize Eskimo anger. The blocked-off areas were huge. Laid over the East Coast, they took up an area starting on Richard's maps 250 miles north of Syracuse, New York, spread south and included all of New York City and Philadelphia, and continued down the coast to Savannah, Georgia, encompassing half the coastline of New Jersey and the entire lengths of Delaware, Virginia and North and South Carolina.

This was a rather extensive area to block off from development, was Richard's point.

A second map showed a federally designated "no-disturbance zone" for bears, in the North Slope, almost totally taking up the barrier islands that protected much of the shore.

"We weren't consulted on this," Itta told Senator Begich when they met in Washington that month. "It's a farce." Itta planned to sue, he said. He wasn't doing this to benefit oil companies, just his people, but if he managed to shrink or reverse the designation it could only benefit Shell and any company seeking to build a pipeline.

Around the table in Begich's office were Rex Rock and Richard from the ASRC; Itta and Andy Mack; Harry Brower Jr. and George Noongwook from the Alaska Eskimo Whaling Commission and Itta's chief lobbyist, Georgetown-based lawyer Alan Mintz, a low-key presence who had worked for decades for North Slope mayors.

Once again Iñupiats were being asked to pay the price for excesses occurring elsewhere on earth. If CO_2 emissions were causing temperatures to rise, they sure weren't coming from factories in Barrow. If ice was melting, and *ice* was what bears needed, how would blocking off huge swaths of *land* increase the amount of ice? In addition, as Itta claimed to Begich, bears were adapting. Sure, they were coming ashore earlier in the year, but hunters insisted the population numbers were high. Official Washington was ignoring these claims, just as hunters had been ignored and turned out to be right, Itta said, in the days when whale population was the issue. In some parts of the North Slope, Itta told Begich, polar bears were mating with grizzlies, producing a new hybrid bear, part polar bear, part grizzly—a huge humpbacked creature observed in 2010.

In short, Itta seemed no happier with national environmental groups than he was with oil companies. Both, given a free hand, would try to control Iñupiat lives, he felt.

Edward and Richard kept up the polar bear argument at the Reagan Center, on their panel.

When it came to development versus preservation, Richard told the audience, "Some people say, You're conflicted. I say heck yeah we're conflicted. Everybody is conflicted. We are all conflicted. We appreciate conflictedness because we know what it means."

Richard good-naturedly made fun of the professional climate-change crowd who show up in the High North in summers.

"Each year after the geese but before the walrus come the climate-change journalists," he said.

Pete Slaiby was also on the panel, and that day, for the first time in public, Itta praised him. "Shell has been more effective than our own federal agencies. We still have concerns but we have been working well together."

In private Itta asked Slaiby for information he was waiting for from the company—Shell's underlying data supporting studies on the air pollution their drillship would release. Itta had been promised this data but had not yet seen it. If he liked it, he'd go to bat for Shell's one-well exploration plan in 2011 with the EPA.

Unfortunately, the information had been forwarded by computer as part of a package and was lost at the moment, Shell said. No, the wrong information had been sent, Andy Mack said. In either case it was a missed opportunity. Itta never did go to bat for Shell that January. Later, when he saw the information, he decided the plan was okay, but he'd already left Washington.

As for Pete Slaiby, he figured he had about a week left before he would have to make up his mind what to do.

CHAPTER 12

What Norway Can Teach the US, January

The blizzard came up swiftly, blotting out stars and driving sheets of snow across the harbor. Hammerfest, Norway's streetlights and ships—visible moments before—disappeared. But a mere inch away from the storm, behind floor-to-ceiling glass, a piped-in piano sonata played softly in the Rica Hotel dining room, reducing the whiteout to backdrop. Guests at candlelit tables breakfasted on a delicious buffet; flavored yogurts and omelets with bacon, thick fresh breads, salamis, hams, cheeses, muesli, waffles, kir, half a dozen kinds of coffees, juices and granola.

Hammerfest sits at a latitude slightly south of Barrow's and it is the site of the Snow White natural gas field, 88 miles offshore beneath the Arctic's Barents Sea. It supplies the Eastern United States. By January 2011, Norway was the world's fifth largest exporter of oil, second when it came to gas. Much of that came from its Arctic.

A new Statoil company bus—normally used to transport workers—pulled up outside the hotel. I climbed on for the ride to Melkøya Island, site of the gas-processing plant and transport docks, in a fjord just offshore. Lavender light seeped into the world as the storm abated. Avalanche barriers held back snow on high ridges ringing town, and homes featured steeply slanted alpine roofs. We passed brand new, modular ski chalet–style town houses. Hammerfest—once home to Vikings—is no stranger to making international energy news. In 1891 it led the world in installing electric lighting.

I was in Norway because when it came to Arctic hydrocarbon extraction, the small Scandinavian nation, population 4.8 million, was the country that all interested parties were talking about in the United States.

The Deepwater Horizon Commission report, out days earlier, called for new offshore exploration and extraction regulations "at least as stringent as those in peer oil-producing nations such as Norway." Even *after* exploration has begun, the report suggested, US "industry should be required to constantly update its risk management plans to reflect actual experience," as in Norway and Britain.

With the commission using Norway to push for new regulations, the logical guess would be that industry resented it, but Shell's top lobbyist Brian Malnak had been filled with praise for the Norwegian system. He called it more cooperative than the US way, less likely to land companies in court, and based on more trust between parties.

And since oil companies saying "trust me" engendered the opposite reaction on the North Slope generally, it was a surprise to learn that Mayor Itta's staffers were also bullish on Norway. After spending months analyzing its oil and gas activities they had sent Itta an inch-thick report.

"Environmental protection is now an integral part of Norway's energy policy and management of its petroleum resources," the report said. "Norway has a strong regulatory framework in place, supported by a proactive industry."

Mayor Itta, sitting in Barrow, read, "Norway's approach to

granting licenses fosters collaboration...A fundamental precondition for petroleum activities on the Norwegian continental shelf is coexistence with other uses of the sea and land areas affected."

The report was not universally positive, citing recent "offshore skirmishes" between fishermen and seismic vessels, and also a 2007 oil spill in the North Sea. But it was undeniably supportive. "The Norwegian government and the oil industry have devoted considerable effort and financial resources to support an ambitious environmental monitoring program evaluating impacts of oil and gas development to the marine environment," it said.

In short, unlike the conflict-ridden United States, Norwegian players cooperated. But how had this happened? Norway—like the US—had fiercely independent indigenous residents, the Sami people, in the Arctic. Its environmental groups opposed offshore extraction. Its commercial fishermen feared oil spills. Norway was a democracy, not a nation where the government arrested critics, as was the case in some other oil-producing countries. So what exactly was Norway doing that kept most parties in balance and *how had this come about?*

First, the tour. The bus descended into a tunnel gouged through rock 70 meters below the fjord. As we emerged onto the island I saw flames venting from a smokestack. It was excess gas—more than the processing system could handle—being burned off. Guards checked my ID. The storm waxed and waned.

"It might be a little cold when we go out," said Øivind Nilsen, facility head, greeting me in his office with another buffet—fresh salmon, eggs, grapes and tangerines and apples—covering his long conference table.

Øivind recited facts. The Norwegian continental shelf held some of the biggest oil and gas reserves in Europe. Hydrocarbons represented a quarter of the country's GNP, half its exports, 75 percent of its trade.

Norway's GDP was $414.46 billion in 2010 according to the International Monetary Fund, as opposed to $14.66 trillion in the United States.

And Snow White was the largest industrial development in northern Norway. Other than the processing plant and docks on the island, everything was underwater. Below the sea, pipes carried the natural gas to shore. There it was chilled to minus 163 degrees Celsius. Propane and butane, separated out, were exported. Carbon dioxide was either pumped back into the well, or, if it was emitted into the atmosphere, taxed by the government.

The main end product, liquefied natural gas (LNG), was loaded onto ships for a twenty-day round trip to Chesapeake Bay in the United States, or a twelve-day journey to Bilbao, Spain. Each ship carried enough LNG to power the city of Amsterdam, near Shell's world headquarters, for six months.

"We have enough gas here to produce at this level until 2048," Øivind quietly bragged.

There had been some software and gas-flow problems, he admitted, but said these were under control. The flow problems occurred if too much gas came through the pipes, and the height of the flames venting from the stacks outside indicated the amount of excess. The constant glow in the winter dark was the new aurora borealis, the new northern lights.

"We do not like to burn the gas. It is gas we could sell." Still, there had been some complaints of soot in town.

We strolled through the $10 billion complex. In the control room a staff of seven monitored 32 screens tracking operations. One man worked full-time on ship loading, as no two ships ever loaded up beside each other. LNG was explosive, and you didn't want to risk a double blast. Another operator worked utility systems, air supply for the factory, compressing units to cool incoming gas. A floor-to-ceiling screen showed all operations simultaneously.

We donned Day-Glo orange body safety suits and steel-toed boots. Outside the wind kicked up again. At two in the afternoon we trudged through what would have been a deep outer space–like Arctic night if not for the glare of bright floodlights on steel pipes, tanks and machinery. The tanks held chemicals for deicing. Immense breakwaters surrounded the complex, levees against Arctic storms. This part of the island was man made. No workers were visible but, "if there is an alarm you will see many," Øivind said.

Øivind proudly added that more Arctic projects would come online soon, even further north. Statoil had partnered with a Russian company to develop Russia's Shtokman gas field and that mega-project would begin operations within a couple of years.

The benefits of extraction were evident. Plunked down in Hammerfest—or even more striking, in Norway's High North regional capital of Tromsø—a Washington, DC, visitor would never imagine that he or she stood above the Arctic Circle. The city was modern and glittering; with new restaurants, and new cars easing through well-plowed streets beneath gently falling snow. There was a quaint village-like atmosphere in the center of town, a mix of historic Scandinavian architecture and modern angular buildings; a university for 9,000 students; a polar institute and an Arctic cathedral that seemed made of glass like a crystal pipe organ; coastal steamers arriving from various Norwegian ports; cross-country ski trails; a polar environmental center and a museum. As in a Currier & Ives print, fat snowflakes drifted onto well-dressed elementary schoolchildren— new rucksacks on their backs—making their way toward class at 8 a.m. as their parents headed for prosperous Arctic jobs.

The big difference was the quality of light—or rather, natural light was absent at 8 a.m. Schoolchildren would not be out beneath streetlights in such darkness in the lower 48 states. They would be home, asleep, because it would be 3 a.m., not 8 a.m., time for school. In the lower 48, this kind of light—man-made—would shine on lonely drinkers weaving home, a delivery truck stopping at a closed supermarket, a police car making late-night rounds.

Fifty years ago—before oil was first found in Norway in the North Sea—the country was far less affluent. By 2011 the national pension fund, from hydrocarbon profits, owned 1 percent of all stocks on earth, Henrik Width, a Norwegian deputy consul general, had told me in New York. It was the largest sovereign wealth fund in the world.

"The recession that hit the rest of the world was barely a blip in Norway," Henrik said.

In January 2011 the krone was the second strongest currency in Europe after the Swiss franc. It was far stronger than the US dollar. When it came to foreign exchange, my hamburger in a modest Tromsø pub, with French fries, cost $38.

Tromsø, Hammerfest and Snow White were impressive, no doubt about it. But the *Deepwater Horizon* had also been impressive thirty seconds before it blew up, during a week when BP officials were celebrating a safety milestone, that workers on the rig had gone seven years without a lost-time accident. So what was it about Norway that made parties in the United States—people who pored over the reasons for the Macondo disaster—praise the system that had allowed the Snow White facility to exist?

A partial answer was provided by Magne Ognedal, Norway's director general of the Petroleum Safety Authority, in an interview he granted to *Bud's Offshore Energy* blog three months before the *Deepwater Horizon* exploded.

"My experience is that you unfortunately often need a major accident or even a disaster to engender political support for streamlining regulatory regimes," he said, not praising *more* regulation but *streamlining the process.*

Magne told the blog that Norway had initially made several mistakes when it came to producing offshore oil and gas, all of which seemed to mirror the way things were done in the United States into 2011.

First, early on Norway employed a single government agency to give out oil licenses and promote safety. As in the United States, this created a conflict of interest. The same regulators were supposed to raise revenue yet slow its flow while it considered safety. Norway later split up those roles.

Otherwise, "it would be like the wolves watching the sheep," one high government official told me in Oslo.

Also, Norway initially used many regulatory agencies to oversee exploration and extraction. This too did not work.

At one time, Magne told the blog, thirteen different agencies regulated oil work. Two accidents changed that. An offshore blowout in 1977 sent 9,000 tons of oil gushing into the North Sea. And the capsizing of a semisubmersible drill rig in 36-foot-high waves killed 123 workers in 1980.

After an investigation, Norway changed the system. By 2011 only three agencies dealt with drilling—one responsible for safety, one for production, one for pollution control and environmental protection.

Magne seemed to describe Shell's problems in Alaska—without mentioning a company—when he told *Bud's Offshore Energy* blog that in several oil-producing countries, "many sets of national acts and regulations include requirements which are confusingly spelled out, incomprehensible, inconsistent and incomplete. Often you will find there is a plethora of requirements that do not provide for—or even concern—safety…Regulators…become seriously bogged down in unimportant details…whereas the bigger issues might be left unattended."

Magne's main concern was safety, while powerful Ole Anders Lindseth dealt with leases and production. We sat down over another breakfast, this one in Oslo. Lindseth was director general of Norway's Ministry of Petroleum and Energy, and when it came to granting offshore leases and monitoring operations, he oversaw the two agencies splitting up those jobs. He filled much of the combined role that in the US Ken Salazar at the Department of the Interior; Jane Lubchenco, who ran NOAA; Michael Bromwich of BOEMRE and Lisa Jackson, director of the EPA, did in the US. Their agencies all—at some point—had to okay a lease or exploration plan before it could go into effect.

Ole was bulky, 61 years old and spoke slowly and with thought. The list of differences between the Norwegian system and the US one was striking from the first. In Norway, he said, companies did not pay for offshore leases but merely laid down a $10,000 application-processing fee. The low cost at the beginning was designed to make things easier for safe production, the eventual goal. There was no complex process of permit applications involved after a lease was granted. The lease *was* the permit. "In one way or another you get to drill," said Ole, because all appropriate analyses had been done *before* the lease was granted. Issuing a lease occurred after a "buildup to consensus" between government agencies, environmentalists, commercial fishermen and residents, Ole said.

For anyone thinking this streamlined system made Norway a patsy for oil companies, Ole added that if a project struck oil or gas, profits were taxed at a rate up to 78 percent, plus a carbon tax.

"No company that I know of has turned down a lease because of the taxes," Henrik Width told me in New York.

In addition, Ole explained, *no single company* was ever awarded a lease. All projects were joint ventures between several companies. The government awarded a lease to a consortium, assigning each company a role. One company did exploration and production and got 40 percent of profits. The others helped bankroll the project and helped oversee work.

"This has two advantages," Ole said. "Everybody is looking over everybody else's shoulders. It's checks and balances when it comes to safety. Also, different companies have different strengths. You may have good geology from company A and good production from B. With five companies you have five chances to find oil. Remember, the operator is spending other companies' money, so the risk is spread."

In Norway, therefore, the process from the first had the kind of clarity that Pete Slaiby and Edward Itta craved. And once operations began, Ole said, safety monitoring was conducted differently from how it's done in the United States, using methods that William Reilly—at an oceans conference in Washington that month—urged America to incorporate.

Norway required a final result, not a specific means of getting there. Companies separately assessed risk for each operation and showed how they would mitigate it, instead of meeting a uniform checklist of requirements. The standard had come about because regulators, after watching the failure of early efforts, realized that rigid rules could actually work against them. Industry technological advancements could outpace rules. Companies, having satisfied outdated rules, could grow complacent and ignore new dangers. Agencies would always lack enough money or manpower for full-time monitoring. So the Norwegians put the onus on the companies. The government gave them the end result desired, and each joint venture had to come up with a plan uniquely suited to that project, submit it and regularly update it.

Did it work? Between 2002 and 2007 not a single fatality marred an offshore oil or gas project in Norway.

"A regulator must have faith in the industry participants' genuine eagerness and willingness to achieve compliance," Magne Ognedal told the blog. "If not, the regulatory system will fall apart as distrust otherwise might easily develop between the regulator and industry. This goes both ways, since a regulator should also be trusted by the industry to do the right things."

In short, ironically, what US players envied in Norway was something they all lacked, trust in the other guy's desire to do the right thing.

Susan Childs was in Norway that week, and so was Adm. David Titley, still planning the US Navy's Arctic Roadmap; and Pablo Clemente-Colón, chief scientist at the National/Naval Ice Center, who had been on the *Healy* in summer 2010.

They all sat in an auditorium in Tromsø, listening to Morten Smelror—director general of Norway's Geological Survey—talk about Arctic diamonds.

"Canada is now producing 15 percent of the world's diamonds," he said. And this from an area once thought to be impossible for diamonds—deep rock beneath the tundra in the once-inaccessible Arctic.

Two decades ago there were no Canadian Arctic diamonds, but by 2011 four mines were open, one owned by De Beers.

"Virtually all of the Arctic will soon be under national jurisdiction," he said, referring to the undersea land claims process going on under the Law of the Sea Treaty. "When it comes to natural resources, 90 percent of the Arctic will be a private lake," he said.

The conference was called "Arctic Tipping Points," and the venue at the university was packed with foreign ministers, Russian and Scandinavian shippers, oil company representatives, Sami people, scientists, environmentalists and journalists from around the world.

"There are piles of nodules on the sea bottom," said Morten, showing slides of various Arctic regions. His talk was titled "Mining in the Arctic: The Heat Is On!"

Click. On screen appeared a map of Baffin Island, Canada. Iron ore from the Mary River mine there will soon be moved by ice-breaking ships, he said.

With 90 percent of rare earth minerals produced currently in China, the Arctic could be the next good source, he added. This would lessen a Chinese near-monopoly on those minerals.

Click. A shot of the Russian Arctic appeared. "The estimated value of Russian northern minerals is between 1.5 and 2 trillion dollars."

Click. "Only 15 percent of Greenland is ice-free at the moment, yet it has shown good potential for gold, diamond and metals."

Click. The North Slope appeared. "Gold extraction is expected to rise in Alaska's Arctic. Alaska also has huge copper reserves in Bristol Bay."

Even if some of the bounty he discussed lay slightly south of the Arctic Circle, the tankers carrying it would, the thought was, soon deliver it by moving north. Norwegian shipper Felix Tschudi and Russian Mikhail Belkin, assistant director of Rosatomflot, which ran Russia's atomic fleet, assured the audience that the great movement of tankers across the top of the planet had begun. In 2010, one of Tschudi's 41,000-ton iron ore ships had made the Arctic journey from Norway to China—through the Northeast Passage—in eight days. Belkin said that Rosatomflot proved in 2010 that vessels weighing 150,000 tons dead weight could make the journey in waters once thought too shallow. A passenger vessel had made the trip from Murmansk on the west coast of Russia to Vladivostok in the east in seven days. A Russian icebreaker had escorted the *Tor Viking II*, the ship that worked for Shell in Dutch Harbor, from the Bering Strait to the island of Novaya Zemlya.

"The window is now open for commercial vessels from the end of June until the end of December," Belkin said.

In fact, Russia had by January 2011 received more than 30

requests to escort commercial vessels by icebreaker across the Arctic that coming summer, he announced.

As for smaller vessels, I ran into several sailors who had been anchored off Barrow the previous September when the *Healy* dropped me off there. They were the crews of two boats that circumnavigated the Arctic in 2010, completing the trip through the Northwest Passage and the Russian Northeast route too.

"Nobody expected us in Barrow. We just showed up," said Russian Elena Solovyeva, who, along with Daniil Gavrilov, had sailed a 45-foot yacht from Russia around the top of the planet that summer.

"In my father's time you could not have done this. There was too much ice," Daniil said.

They'd stopped in Barrow, they said, to sit out a gale and buy supplies. Going ashore, they were befriended by locals. But they were surprised because they never went through customs. There weren't any customs agents in Barrow.

The same thing happened to Norwegian Børge Ousland, who started his Arctic circumnavigation on June 12, 2010, in Oslo and returned on October 14 that year. The 48-year-old self-described professional polar explorer had already skied solo across Antarctica when he and Thorleif Thorleifsson completed their sailing journey in a 31-foot fiberglass trimaran with a ten-horsepower motor, the first Norwegians to sail the Northwest passage since Roald Amundsen did it in 1908, and "it took Amundsen almost six years to get through," Børge said.

"Even six years ago my trip would not have been possible," said Ousland. "Because of ice."

His stop in Barrow, he added, "was a highlight because that was when we'd crossed the Northeast Passage."

In Barrow, "no one asked us any questions." Security-wise, he said with a grin, Barrow was "the open keyhole where you don't see a key."

Børge was cordial, and no stranger to being interviewed, part of a modern professional polar explorer's job. He said one reason he chose the trimaran was maneuverability. In icy waters, sometimes one runner stayed in water while the other could move over the ice like a sled.

"But I was surprised because we saw very little ice. And the bigger surprise was the lack of animal life when there was no ice. When in the ice we saw all sorts of seals, walrus and birds. But when ice was gone it was like a desert."

Weather-wise, Tromsø had advantages over Barrow when it came to development, as it was warmed by the waters of the Gulf Stream. Temperatures there were more like Anchorage's, averaging a year round 36 degrees. There was no permafrost. These more comfortable conditions also made it easier to house the headquarters of the Arctic Council in Tromsø. The council is the permanent, informal body that functions as a mini UN for polar nations, a center for mutual decisions made about the entire region.

In the council, by 2011, diplomats from member nations were trying to work out Arctic shipping and safety rules that would apply across the region, a "voluntary code" for shippers. Council representatives also signed a historic search-and-rescue treaty that year, designed to expedite speedy aid for ships in trouble. It laid out spheres of responsibility—who does what and where—when future Arctic sea disasters, expected by all, occur.

Like Barrow, Tromsø boasted a polar research center, where scientists from more than 60 countries study Arctic issues. It was a nexus for global research on the north.

But unlike Barrow, Tromsø played an important role in the nation's high priority Arctic strategy, its focus on the region as critical to the future. With a third of Norway lying north of the Arctic

Circle, with 10 percent of its people living there, and with the Arctic part directly attached to the rest of the country, unlike Alaska and the United States, it was easy to see why even average Norwegians thought it important to address issues in the region.

There were plans in Tromsø to develop Arctic cod farming and encourage outer space–related industries—satellite companies—to develop overhead aids to Arctic navigation.

"Everyone expects that there will be an increasing need for updated information for sea traffic security," said Jan Petter Pederson, vice president of Tromsø-based Kongsberg Satellite Services. The company had initiated a service for detecting and monitoring icebergs.

Tromsø was also where Norway's biotech and bio-prospecting companies were establishing themselves and seeking medical miracles—natural substances to fight diseases—in the Arctic. One subsidiary of Tromsø-based Biotec Pharmacon had developed an enzyme from codfish that is now used in commercial HIV tests.

In the future—in the same way that new drugs have come from more moderate climates—a childhood leukemia cure from the rosy periwinkle plant of Madagascar, anesthesia from curare, a poison from a South American fish; an ovarian cancer drug from the yew tree—researchers are beginning to seek medical answers in the High North.

In Barrow, for instance, staff at the borough wildlife management department told me that they've received calls from medical researchers in New England wondering why bowhead whales can live for over 200 years. Perhaps there would be a longevity secret for humans in the DNA of bowheads, the scientists thought.

The North Slope researchers had begun to get discreet inquiries from scientists seeking samples of bowheads from hunts. The dispensation of whale samples was prohibited. Welcome to new biomedical ethics problems in the High North.

Jonas Gahr Støre, Norway's foreign minister, was a slim man dressed in black, with grayish hair, an expressive face and the air of a focused, stylish pastor. His white cuffs stretched from well-cut jacket sleeves as he moved his hands to make points. We met in a Tromsø pub over a snack of spicy, delicious codfish stew.

"The Arctic is new geography, new geopolitics...We see the rest of the world casting their eye on the Arctic and we are at the doorstep," Store said.

Store's assessment of the Arctic political climate matched ones from the US State Department. He believed northern countries were cooperating and that through the Arctic Council their diplomats were working out step-by-step processes for co-managing and protecting the north. He praised Norway's neighbor Russia but was also wary of it.

"Russia is inconsistent."

Once again the huge nation was the wild card. Russian policy and internal politics, he felt, and the way other nations reacted to them "will be one of the main drivers of what happens in the Arctic," Jonas Gahr Store said.

So much came back to Russia. In the same way that the Soviet Union started the space race in 1957 by launching *Sputnik*, the first satellite, Russia in 2011 led the push into the Arctic. The difference when it came to US reaction was that when the space race began, a prosperous America had been able to start out second in a competition, bypass the Soviets and come out ahead. But in 2011 the United States was saddled by debt, its populace oblivious to a new challenge from an old rival. Even if America decided that the Arctic was crucial to national interest, it was unclear if it could mount the effort required to gain primacy there.

"Norway has been at peace with Russia for a thousand years,"

Jonas Gahr Støre said, implying, as diplomats usually did at first, that the Arctic would continue to remain as peaceful as it was in January 2011. But then he added the usual caveat. "Russia is a country going through immense changes. We don't know what state Russia will be when that transition is completed. A vulnerable democracy. There are grave questions to be asked about corruption and rule of law."

He spoke about more than just security. On the environmental end, with climate across the rest of the world linked to Arctic conditions, with southern fish stocks moving north, and earth's common northern bounty endangered by warmer weather, no matter what caused it, he said, "It's one ocean up here. We need common standards. Unless we have them we will run into trouble."

Bottom line? Jonas Gahr Støre uttered the same words that whale and walrus hunters had used on St. Lawrence Island and in Barrow, that Mayor Itta had employed, that the US secretary of homeland security had spoken on the bridge of the *Healy* one 6 a.m. two years before.

"There's less ice."

I took a brand new, silent-running tram from downtown Oslo to the clean and new-looking airport. In the Continental Airlines jet on the way home, just as when I flew to New York from the North Slope, there was the sense of traveling a route corresponding to an immense straw sucking hydrocarbons from the earth, transporting it to the US lower 48 states. Oil-wise, the United States remained in 2011 the world's biggest per capita energy terminus. China had overtaken America as the nation consuming the most energy due to its mammoth population, but as individuals, US citizens were earth's gluttons—their cars bigger, their homes heated higher in winter, their computers and TV sets and iPods running around the clock.

Traveling eastward from Barrow to New York, when skies were clear, I regularly looked down on homes growing more closely packed, roads more congested, towns obscured by thicker smog, a visible appetite for power that became—the further east we flew—more extreme.

It would be nice to think that clean alternative forms of energy—solar or wind, hydrogen or nuclear—might easily take up the slack if oil production declined on earth, during any interim period separating the current oil era and whatever energy source will next dominate the globe.

But the fact was that no current energy source could accomplish this in January 2011. When Pete Slaiby or Mayor Itta or environmentalist lawyer Peter Van Tuyn came to Washington, they all needed oil to fuel the planes that flew them to the capital. At home they needed it to power the cars they drove. Whether Shell explored for oil in offshore Alaska or not in 2011, hydrocarbon needs in the United States would keep rising. The nation was filled with energy junkies who had no intention of kicking the habit, and even if they did, earth's population and power needs kept rising. If the oil didn't come from the North Slope, it would come from Iran, Venezuela, Russia or Norway. The exhaust would still go into the air and the profits from it would go elsewhere.

The plane landed and I took a taxi home toward Manhattan, where, in winter, smoke rose from high-rises burning oil, and exhaust rose from trucks idling on Sixth Avenue. Movie marquees sucked up energy. The factories that produced the coats people wore on the street were powered by oil, coal, gas. The restaurants I passed were warmed by oil or gas. Stop oil and the nation and the world would go dark, physically and emotionally.

My own apartment building was heated by oil, and walking in, I appreciated the warmth.

The phone rang shortly after I got home. Shell's Curtis Smith was calling from Alaska.

"We're pulling the plug on 2011," he said. "You know. The EPA."

CHAPTER 13

A Bittersweet Goodbye, February

The drums were made of whale-liver lining stretched over steamed hardwood. The drumsticks were flexible woods. Some songs were a thousand years old. Celebrants had been arriving in Barrow for days by plane and snowmobile. Land travelers could often be identified by white patches, frostbite, on their faces. The one-room terminal at Wiley Post Airport—luggage bin on one side, bathrooms on the other—crowded with boisterous people each time a flight arrived. SUVs idled outside, their engines running to keep interiors warm. Homes were packed with visiting relatives and friends sleeping on couches or floors.

"This *kivgiq* will be the last one for Mayor Edward Itta."

The traditional celebration—"messenger feast"—began on February 9, when Edward had eight months remaining in office. He would not run again. North Slope mayors cannot seek three consecutive terms. By now it was official that no offshore oil exploration would occur in America's Arctic during the rest of his time as mayor.

Beginning right after Shell pulled the plug on hopes for 2011, the four-day event in Edward's honor had been planned for months by the Borough Assembly. It would feature dancing, singing and gift giving. Assembly president Eugene Brower, grandson of Yankee whaler Charles Brower, kicked things off in the high school gym.

"We want to thank the mayor for six years of service to the North Slope."

Like Olympic athletes, drummers and dancers paraded into the gym, grouped by village, singing loudly and beating handheld drums. Smaller villages sent only a few dancers. Barrow had the biggest contingent, as host community and largest on the North Slope. Banners strung from the ceiling announced the Kotzebue Northern Lights Dancers and the Kaktovik Dance Group, the Utuqqagmiut Dancers, the Tagiugmiut Dancers, the Suurimaanitchuat Dancers.

"The mayor has been a strong champion for us in the state, nationally and internationally. The mayor was able to see the big picture."

The next election would be in October. No candidates had identified themselves yet, but within a year Edward's replacement would be the one meeting with White House officials, the secretary of the interior, EPA, Navy and State Department people and any petroleum company representatives seeking to do business in America's Arctic. Many waited in the wings.

The gym reverberated with drums. *Boom...boom...*

"The mayor paid attention to the small details."

Outside, temperatures ranged between 10 below and 50 below zero. Winds gusted to 30 miles an hour or dropped to dead calm. An *ivu*, a wall of ice, had risen up offshore, slid onto the beach and stopped before the coast road to loom above passing cars like a watching animal. Would it come closer? *Did you see the ivu*, people asked on the street, in restaurants, in school.

"Mayor Itta's work does not only impact the North Slope. It

impacts other communities who do not have the money to fight the federal government and the corporations," Eugene Brower announced.

There would be fireworks and fiddling, storytelling and skits. All around town hunters climbed down ladders to their permafrost cellars, stood appraisingly amid the sheen of iced-over whale and caribou meat, and hauled glistening hunks topside to feed the multitude.

A *kivgiq* celebrates winter's darkness ending, clarity's seeping back into the world. It is a welcome break in the tedium of Arctic night. Messenger feasts were observed long before oil was discovered in Alaska or automobiles existed or there was an entity called the United States. *Kivgiq* messengers—runners summoning guests to feasts—spread across the tundra on foot and by dogsled before Russian explorers claimed Alaska, before Russia sold Alaska to the United States without the Iñupiats knowing they'd been "purchased" to start with or that, as in any real estate transaction, papers had been signed, money had changed hands and whole villages had acquired new "owners."

Festivities began with a race of high school students from the airport to the gym in minus-twenty temperatures. Mukluks pounded on the basketball court when the kids burst in to wild applause. The winning boy, teamed with an elder, lit a seal-oil lamp to commence one of the northernmost celebrations on earth.

Edward looked happy, sitting beside Elsie, in matching white parkas. They occupied fourth-row seats, as the first rows of folding chairs were reserved for true VIPs, which meant elders. Itta was as effusive as any mayor, greeting people, making jokes.

Called to the podium, Edward announced, "This will be an alcohol- and drug-free celebration."

The applause was loud.

"Of all the things that God could have made me, I thank God that I am an Iñupiat," Edward said emotionally.

The applause grew huge. Almost 800 people, 10 percent of his constituents, shared in the happiness.

But later in his office, when he spoke of Pete Slaiby's phone call to him days before, he did not smile at all.

The call had come one day at 4 p.m. and in it the head of Shell's Alaska Venture told the mayor there was no way Shell's *Discoverer* would drop anchors off the North Slope in 2011.

Slaiby described his last-ditch efforts to keep Shell's plan alive; the trip to Washington in mid-January, the meeting at EPA headquarters with deputy administrator Gina McCarthy, the odd sense in that room of finding himself allied with federal agency heads against their own review board as they tried to "determine if there was any heroic effort the EPA could make to get the permit to work in 2011."

But the remand had come down too late.

"Because after they fix the permit it will have to undergo another public comment period, then another period of incorporating those new comments into the plan, then another trip back to the appeals board," Slaiby said.

Proceeding with the plan in 2011—Slaiby said—if Shell chose to try it, would mean once again gambling with tens of millions of dollars to lease equipment *just in case* the remand process sped up and ended in Shell's favor by summer. Had this been the first year the company faced that choice it might have gambled. But after four years of losing dice throws, by the end of the Washington trip Pete knew there was no way to proceed.

So Slaiby reluctantly phoned Dave Lawrence in Houston and advised Dave to pull the plug, he told Itta. Dave concurred. The public announcement—the admission of failure—was tricky because it could cause Shell's stock to drop. It would be made by Shell's CEO, Peter Voser, in the Hague, Slaiby told Itta. Voser would blame US federal policies and permitting decisions for the delays.

"We will try again in 2012."

Edward didn't care who made the announcement. He cared that the drilling had been blocked, and in that phone call—for the first time since he'd started fighting Shell—he found himself surprised.

"I was disappointed."

Alone in his office, phone in hand—Edward asked himself if he had pushed too far this time, tried to wring out one too many concessions, failed to weigh in adequately for a plan that had been changed to accommodate his objections and that he had come to approve of.

Sitting near reminders on his wall of what a good Iñupiat leader should be—the little painting he'd hung of the hunter trapped on an ice floe because he did not plan properly; the larger oil painting of borough founder Eben Hopson, the visionary who had helped create a political entity to wring tax monies from oil companies...funds to pay for schools, plumbing, heating systems and rescue squads—Itta had felt the weight of history, culture, commerce and choice.

The disappointment deepened into a horrible sensation of failure.

"I wondered, Could I have done more?"

He'd been prepared to go to the EPA and argue for the Beaufort plan, but the supporting emissions data from Shell, due to misunderstandings between staffs, "didn't come in time for me to say there was no impact on our communities."

The bottom line in February was, "I'm more comfortable with Shell. They are making a sincere effort. It's like I told Pete in that meeting last spring. Don't leave it to me to explain things. *You* convince our people."

He was quiet a moment and in the silence I remembered our first meeting in which I'd asked a favor. I'd told him that I knew that Iñupiats were taught from childhood to downplay the individual and elevate the group. That usually North Slopers shied away from talking about "me" and described events from "our" point of view. I had asked Itta, even though it might be difficult for him, if he would mind talking about himself over the next year, *his* thoughts, *his* hopes, *his* history.

Now Edward said, "I think about this all the time. I try not to dwell on it. But I keep coming back to needing to maintain the economic well-being of the borough so my children and grandchildren have something later on. That's why I tried so hard to elevate standards so we can try to do any oil exploration offshore as safely as possible...

"But the other scenario is, *no* production happens and our economic base goes down. With no oil in the pipeline we'd be back to 40 years ago. I shudder at the difficulty that our people would face."

Itta said he had found himself after hanging up with Pete Slaiby remembering his uncle George, who had taken him hunting when he was a boy and understood even before any oil was discovered on the North Slope that big changes were coming. But George, unable to adapt to the changes, had killed himself in one of the worst events of Edward's life.

"Time has taken the pain away, but when that call from Pete came I missed my uncle terribly. I'm thankful that I had an uncle like that. He would tell me when I goofed up," Edward said, voice beginning to falter. "He would spare whupping me. That was left to my papa. But he was an influence on who I am today."

Pete Slaiby's call had conjured up his uncle.

"The effort I made. The disappointment. I felt that maybe I had not done enough." The glint of moisture appeared in his eyes. "How could I have done better? As mayor only I must live with that.

"It's never been easy for me to talk about myself," he said. "Maybe I take this too heavily. But when I get discouraged, it is there. My uncle told me, 'You can cry, but don't give up.' My uncle told me, 'You are on the right track.'

"My uncle," Edward said, "is there all the time."

At Shell that month, in more casual conversation, I sometimes heard people speculating about where the company might send them next when their time was up in the High North. The Mideast? New Orleans? Their frustration over the offshore delays was genuine, but many of them had not been born in Alaska; they knew that the company rotated people to postings and expected at some point to leave. Over drinks with Curtis and his wife, Jody, the talk turned to where they might be assigned next, and to when Slaiby's turn to leave might occur.

Folks in the Anchorage office might be dealing with Arab sheiks within a year or two instead of with a North Slope mayor. They might be in Norway or any of a hundred other places on earth where Shell had rigs. These people were dedicated to their careers. They were hard workers. They took results to heart. But in the end they would leave.

Mayor Itta was anchored to the North Slope, and the consequences of drilling—good or bad—would be part of his life and his community and his nights forever.

That week above the North Slope the sky produced one of the Arctic's greatest shows on earth, the greenish streaks of aurora borealis, the northern lights, obscured in the heart of town because streetlights blocked it out, slightly more visible at BASC, farther away, occasionally sharp and glorious at 3 a.m. at the northern tip of the coast road—America's last ten feet of wind-blown municipal byway.

Boom...boom...boom...

Inside the gym, Richard Glenn and the Barrow dancers wore powder blue and groups took turns playing. The drummers always sat in a single row behind the dancers.

John Hopson Jr.—who had been on the Dutch Harbor trip—wore black like other Wainwright performers.

Older audience members wore their best parkas, of thick blue corduroy or cotton duck outer liner, quilted inner liner, wolf or wolverine ruffs for men. Women's parkas were more colorful. The trick was in the hood. It was not an afterthought, as in lower 48 parkas, but integral to the design. Many men also wore sealskin vests. Older women preferred long floral-motif dresses called snowshirts. Infants were carried on the backs of their mothers. Teens and young parents showed up in jeans, sweaters, Boston Red Sox caps, factory-made boots. The proud sports banners overhead could have hung in any town in Alabama, Oregon, Indiana. WESTERN CONFERENCE CHAMPIONS, 2008–2009, WESTERN CONFERENCE CHAMPIONS, 2009–2010 GIRLS BASKETBALL.

Offshore oil squabbles were not part of *kivgiq*, but critics of Edward's policies were present to take part in the general happiness, whether or not they chose to agree with the mayor. Anthony Edwardson—president of the Ukpeaġvik Iñupiat Corporation and Thomas Olemaun, Native Village of Barrow executive director, who both opposed offshore oil extraction, both gave welcome speeches. Olemaun had his handsome young son along.

Dancing began each morning and lasted well into night. The steady, synchronized drumbeat vibrated through the first floor of the school and with each day goodwill grew. Strangers greeted one another. Old friends exchanged gifts purchased at the craft fair. In the stands sat Harry Brower Jr. and also Taqulik Hepa, head of the Department of Wildlife Management; Geoff Carroll, the musher and biologist; Fran Tate, owner of Pepe's; and Itta's top science advisers, Robert Suydam and Craig George. George Neokok, a native observer I had met on the *Healy*, preferred the top row. We renewed the discussion of food begun three years earlier on the bridge. George still salivated at the thought of fresh clams cut from a walrus's stomach. I preferred lasagna, but it made him sick. We laughed about food preferences and he said that come July, he'd probably be back on the *Healy*, scanning the seas for whales, seals, walrus, and writing his reports for the borough.

Out on the gym floor, audience members often joined in dances, everyone invited to participate at the end. Sometimes before a dance began they were summoned by groups. "Thirty-year-old group!" "Whaling captains' wives!"

When the 50-year-olds' group was called I invented the New York City taxicab dance. I figured that whatever moves I made would look pathetic so I might as well try my own. I explained to the beaming mayor afterward that my jerky flailing and stamping amid the crowd represented hailing a cab, watching it pass in the rain and kicking the curb in frustration.

The whole point was to join in the fun.

Itta's staffers performed a dance in the mayor's honor.

The magistrate was dancing. Three- and four-year-old children darted between stomping dancers, imitating adult moves and animal cries. They became little seals.

Awk!... awk!

"What does that dance represent?" I asked Nok Acker, from BASC, as we watched.

"Lovers traveling on the tundra."

"And that one?" I asked Richard Glenn.

"That's the stewardess dance."

"Excuse me?"

"The dancer is acting out the safety demonstration. See? She's showing how to put on the seat belt. Pointing out emergency lighting on the floor."

Boom...boom...

Assemblyman Mike Aamodt performed a personal gift dance and gave his wife, Patsy, a present. Each noon, shipments of free food arrived, paid for by the borough—boxes filled with turkey sandwiches, apples, cookies and candy bars. A first-floor lounge turned into a kitchen, where volunteers laid out an assortment including whale meat and caribou stew. Diners filled paper plates and sat at any table and chatted with whoever was there. The bowhead meat looked like pot roast and tasted like it too, with a vague fishy tinge. Muktuk, combined blubber and skin, looked candy pink on the fat side, rubbery dark for skin. There were soft drinks and coffee. High school students on break stopped by to eat, and when school was over many joined the audience.

Kivgiq started at 10 a.m. each day and went on until midnight or 1 a.m. Nights brought community dinners at the elementary school or church, where anyone could grab a plate, walk down the line and fill up.

Throughout it all there was no Shell Oil dance, nobody miming paddling an *umiaq* away from a drillship, or a bowhead diverted by seismic booms heading farther out to sea, or a proud Iñupiat working a new oil rig job. But civics and history lessons were embedded in *kivgiq*'s fabric. A man strode to the podium and began speaking earnestly in the Iñupiaq language. The audience craned forward as the speaker uttered two English words, "white explorer," and went back to Iñupiaq.

I figured he was recounting history of the North Slope, perhaps telling a tale of a British ship smashed in ice, or of Yankee whalers, or perhaps a story of whites bringing influenza, alcohol, processed sugar or Bureau of Indian Affairs schools to the Slope.

I asked an elder what the speaker was saying, and sure enough, the words went to the heart of changes between the old and new North Slope.

"Someone left a white Ford Explorer in the parking lot with its lights on," the elder said.

Mark Ivey usually lived in Albuquerque, New Mexico, where he worked for Sandia National Laboratories, operated by Lockheed Martin for the US Department of Energy (DOE). He was tall, youthful despite white hair, and dressed in an impossibly light-weight looking fawn-colored jacket on a minus-twenty degree day. In a rented truck we headed out from the Naval Arctic Research Lab campus onto the tundra, toward two small buildings perched in the distance. Sandia leased a small duplex for an office and for housing researchers or technicians on the old base. The comfort-able adjoining apartments contained several bedrooms with bunk beds, two kitchens—one on each side—cable TV, outdoor gear hanging on pegs in a shared foyer, Internet access, lots of polar bear warnings on walls, and a large thermometer outside so you could open the door each morning and see a red arrow hovering between, that February, 5 and 35 below zero, not including wind.

The arrow pointed toward the parkas and lined overalls hang-ing in the foyer that day and the sun shone brightly at first. We headed toward the old early radar warning system, DEW Line, now automated, approximating an area designated by Sandia and by NOAA—sister research organizations in town, along with USGS—as one of a handful of locations on earth where they gather baseline information on atmospheric gasses. The work enabled researchers to better understand changes that have brought so much attention to the Arctic and their ripple effect on weather and people across the globe.

"Whatever is going to happen in the rest of the world happens first and to the greatest extent in the Arctic," Dan Endres, who had run Barrow's NOAA atmospheric baseline observatory for 25 years, until 2009, told me before I arrived.

The observatory was established in 1973 and was open to researchers from around the world, and by 2011 it hosted more than a dozen cooperative programs with universities and other federal agencies. Instruments at the facility also conducted over 200 routine measurements of air each day, the results shared with science stations operated by other countries—Greenland, Russia, Canada, Sweden, Finland and Norway.

Other sites where NOAA conducted baseline atmospheric science were Trinidad Head, California; Mauna Loa, Hawaii; Cape Matatula in American Samoa; and at the South Pole.

As in these other stations, sensors in Barrow sniffed the air for ozone, carbon dioxide and methane. They detected pollution coming from Chinese factories thousands of miles away. The Arctic is a huge repository for CO_2, a major greenhouse gas. In summers, much of the world's carbon dioxide is absorbed by boreal forests in Russia and Canada. Each autumn, as vegetation dies, much CO_2 is released into the air. This oscillation—the largest fluctuation on earth—has been likened to the planet's breathing.

Based on measurements taken at Barrow and sister locations, scientists knew by 2011 that the yearly average of carbon dioxide in the atmosphere had increased in the Arctic 16 percent between 1974 and 2008 and that methane—the same frozen gas that had been brought up from the sea bottom on the *Healy*—had increased 5 percent between 1978 and 2008, Russ Schnell, deputy director of NOAA's global monitoring division, had told me before the trip.

Correspondingly not only did Arctic ice diminish rapidly during those years but the melt began earlier each year—roughly nine days sooner in 2010 than it had in the 1970s.

The ripple effect from *less ice*—the switch from a heat-repelling white surface of earth (ice and snow seen from space) to a heat-absorbing black one (ocean from space)—accentuated warming across the planet, whatever its original cause—and the results extended south and to the farmlands of the Midwest and the great cities of London, New York, Moscow, Beijing, exacerbating droughts and floods, heat waves and windstorms.

"A small change in the temperature in the Arctic can produce greater changes than in the lower latitudes," Dan Endres told me.

"The Arctic is the mirror of the world."

Barrow was chosen to study that mirror because, "we wanted places far removed from large industrial sources of gasses, yet not so remote that they're impossible to get to," Endres said.

Now Mark Ivey pulled up to NOAA's Barrow observatory and the Department of Energy's ARM (atmospheric radiation measurement) site, two buildings about 25 yards apart that provided scientists with "a front-row seat on climate change," Mark said.

"The Arctic is the canary in the coal mine when it comes to what will happen next on earth."

Outside, a team of hard-hatted workers operated a large screwlike auger in subzero temperatures to grind into permafrost, preparing the ground for installation of more machinery and, surprisingly, *cooling devices* to help keep the ground solid. The pumps—like ones in town abutting the airport and office buildings—would circulate refrigerants to suck away man-made heat that could otherwise buckle permafrost beneath buildings, or in this case research facilities, causing them to sink.

"A main long-term purpose of our work," Ivey said as we climbed steel stairs to the ARM building, "is to use the information in climate models, the big ones that look 100 years into the future."

Inside, it was warm and cramped from computer screens, file cabinets, radars, clocks showing different time zones. The ceiling was a mass of wiring and foam weatherproofing. There was an ever-present hum of machinery. Outside, luminous snow stretched beneath a blue-gray sky.

Ivey said that all major computer models currently predicting earth's future climate said that it would continue to warm, but they disagreed on the rate sometimes. A big goal of ARM's research—funded by DOE's Office of Science—was to examine influences accounting for these differences. By fine-tuning the models, the hope was that predictions coming from them would align more precisely and help Arctic nations make better choices.

Ivey showed off some key instruments.

Millimeter cloud radar sent up radio waves, which, returning, showed how clouds were traveling, at what heights, and how fast particles inside the clouds moved. It provided a sort of X-ray of moisture, important because one disagreement between scientists in 2011 concerned the exact role that clouds play when it comes to heating or cooling the earth. Do they allow the sun's heat to be absorbed by the planet, thereby accelerating warming, or do they reflect it back to space, contributing more to cooling? Or do they at different times do both?

"Right now in most cases computer models don't accurately include the physics of clouds, so our work could reduce uncertainties associated with different models since clouds are the biggest source of that uncertainty."

LIDAR, an acronym for Light, Detection, and Ranging, was similar to radar but "uses a laser pulse instead of radio pulse," Mark said, eyeing a green steady light emanating up. "Actually, it pulses so quickly, a couple of thousand times a second, that you don't see the pulsing. It gives us information on how light is polarized when it comes back...From that we can learn if we are looking at ice or water inside the cloud. That's important for modelers because the light coming from this instrument interacts with ice particles in clouds the same way that light from the sun interacts with them. So if you want to run a model to determine how much light reaches the ground or reflects back to space, understanding that property is important."

Moving on to the *microwave radiometer*, Mark indicated a computer registering the amount of radiation in different frequency bands. "It helps us infer how much water is in the sky."

Politics was irrelevant to the instruments. They would register the same data whether a Democrat or a Republican occupied the White House. To the machinery's steady humming, instead of the beating of drums, one might imagine an Iñupiat dance of the year 2030. Perhaps it will depict whale hunters at sea, shading their eyes as they follow military jets overhead, or gazing at the superstructure of a passing oil tanker or freighter.

Perhaps a new "Visiting VIP Dance" will show chattering diplomats talking while a hunter tries to find a whale in an iceless sea, but with ice gone, the whale is passing to the north, too far away to reach, or to hunt, so there is no whale meat at the next *kivgiq*.

This is the story of *kivgiq*, as told by Martha Stackhouse to the hushed crowd:

This comes from the Coleville River Legend.

A long time ago a man and a woman lived by themselves. All they did was hunt in the wild and butcher animals and put food away. Work, work, work. They were lonely. Then the woman got pregnant, and had a son. He grew up and went hunting and disappeared. His parents were distraught and went looking for him. They never found him.

They had a second son and he went out also and disappeared. They never found him either.

Then they had a third son, a really intelligent little boy. He promised never to stay out at night when he went hunting. But one day when he was hunting he saw something in the distance, a dot that got bigger, and closer, and landed in front of him. It was an eagle and the eagle put his beak up and instantly became human.

"I killed your two brothers because they refused to sing and dance. I want you to come to my eagle mother and learn these things," the eagle said.

The boy did not know what singing and dancing were. But he said okay. The eagle flew him into the sky, to the highest peak. He was so afraid that he did not want to look down. But he did and saw a little sod house.

As they landed the boy heard, Boom. Boom. Boom. He said, "What is that?"

"That is the heartbeat of my eagle mother," the eagle replied.

They went into the hut and saw a decrepit old lady who could not even sit up. But she told them how to make a holiday. How to make drums. The wood. How to use the animal membranes or the liver of a walrus. They made a drum and started drumming. The boy thought, Wow, it sounds just like the eagle mother's heart.

Then the eagle mother showed him how to dance. "Men, dancing, show how strong they are. Women, like birds, look gracefully down from flying," she said.

And then they composed songs together and she said, "This is not enough, you have to give gifts," and they made beautiful mukluks and fur parkas, arrows, leggings...

This all happened so long ago that there were very few people around, so the boy asked the eagle mother, "Where will all the people come from to dance and receive these gifts?" The eagle mother said, "Don't worry, when the time comes they will come two by two..."

After three years the eagle mother told the boy, "You can go home now. Show people what you have learned so they won't be lonely anymore."

The eagle flew the boy back and put him down near his parents' home. He started running, crying out, "Father, Mother, I'm home!"

They ran to each other, hugging, crying, talking. The boy told them about singing and dancing but they did not know about it. They thought maybe he had gone crazy. But they agreed to learn to sing and dance, and they began hunting so that they could make gifts. Every night they would make things; kayaks, umiaqs, clothing of all kinds for gifts, and after another three years of this it was the time of the year when the sun first comes up, reappears after winter. Lo and behold, two people came to the home, and then another pair, and another! They were dressed in different animal skins—polar bears, foxes.

They had a big feast and they brought out the drums. They started singing and dancing. They were amazed. They had such a great time and they gave out the gifts to all the people. They danced all night long.

Then, early in the next morning...we didn't have tall doors then, but little doors. You had to get on your hands and knees to get out...the next morning the guests left and the two people dressed like caribou turned back into caribou, and the people dressed like polar bears turned back into polar bears. You see, there were so few people in the world then that the animals had come to help.

Later the boy went back to visit the eagle mother. When he saw her, she was younger. She told him, "Every time there is

singing and dancing I grow a little younger." And so we still
celebrate and the eagle mother is still alive.

In the gym, as the last day approached, it was time to surprise
Edward with a special gift. A large oil painting was unveiled and
the audience admiringly went "Ooooooh."

It was the same kind of painting that hung in his office and
the Iñupiat Heritage Center and ASRC building—the ones that
depicted borough founder Eben Hopson, Eben looking into the
distance through his black-framed glasses, the bounty of the North
Slope, a whale, sunrise, birds, ice, glistening around him.

In the new painting it was Edward and Elsie occupying the center. They gazed away, also surrounded by the natural wonders with
which they had grown up. Curving bowhead bones arced above
them in the shape of a rising cathedral or swords held by a color
guard to honor a Marine. The background included glorious sunrise colors. To receive a portrait like this in Barrow was a local
equivalent of being presented with—during American Revolutionary times—a commemorative painting made by Gilbert Stuart,
portraitist for George Washington.

Edward received a standing ovation.

"Edward truly epitomizes what an Iñupiat leader should be, and
that is humble," Rex Rock told the crowd as a stream of accolades
from assembly members began.

It was an emotional day. Flanking Edward sat two of his old-
est friends, the boys who had appeared with him in the National
Council of Churches film *Adventures with North American Neighbors* back in the 1950s, when no appreciable oil had been discovered in the North Slope yet, and kids went sledding by sliding
down hills in their jackets, and Harry Brower Jr. played with bone
bits as toy cars. Mark Wartes was now a grandfather and lived in Fairbanks. Lloyd Nageak lived across from the police station in an old

house with a bungee cord instead of a front doorknob holding the door shut.

The Three Musketeers were still together and talked regularly.

"Edward still comes to my house when something bothers him," Lloyd said.

Next came a skit honoring the mayor's Healthy Communities initiative, performed by teenagers from several villages. The drums beat slowly and in the dance, a girl appeared with a boy who wore a cassock with a cross on it, representing religion and stability. But they separated and one by one other teens—temptations—gyrated toward the girl. A boy threw cash at her feet. A girl in a slinky dress mocked her plain clothing. Another girl mimed drinking alcohol and offered her a bottle, the temptations nonstop, and finally the lead dancer began to succumb. Her dance became lewd and drunk and she grew dizzy and sick. And then, at the low point, a boy danced up to her and handed her a revolver. The drums kept beating. She pressed the gun to her head.

This skit could have been presented at any high school in the country. The message was universal but this particular audience seemed to feel it especially powerfully. People began weeping. They barely breathed. A man beside me cried silently into his palms as he covered his face. Every extended family was affected by these problems.

Then on the dance floor the boy in the cassock reappeared and broke through the cordon of teenagers around the suicidal girl, pushed them back and shielded her against their last assault, the bad influences trying to reach her in waves, but breaking against her defender as the ocean breaks against a rock.

Thunderous applause erupted. The girl was safe. *This* was the issue on which Edward had originally run for office, not oil. The teens were thanking him. But they represented only a small percentage of North Slope youth, and Edward lamented—distracted by the oil questions—that he had not had time to do more.

Oil issues still took up so much of his time.

In Anchorage, Pete Slaiby was not sleeping well that month. Late at night as Rejani and Teddy dozed, he'd lie awake tossing, get out of bed so as not to disturb his wife and go up to his study and flip open a book about World War II in Borneo—and US airmen who had crashed there but survived in a hostile, far-off place.

The far-off place might have beaten them, but it didn't. This remained Pete's hope for Alaska too.

Pete had been "shocked" at the extent of the EPA review board's remand. He'd expected that some revision of the Shell plan might be called for but not so much.

And even after the remand came down, for a couple of weeks in early January, "You hope upon hope that there would be a way to make 2011 doable," he said.

"I used to think that one particular point or another in the permit application was key. That if we did *this thing*, or *that thing*, we'd get the permit. Now I saw we were well beyond that. The point was that the EPA has difficulty navigating the issues. They seem to be incapable of providing the guidance we ask for, to tell us what we need to do. By now this has been going on for years.

"Look, nobody at Shell has a problem with strict regulations. In fact it's anybody's guess if regulations in Norway are even tougher than they are here. What is different in Norway is that you don't have this bevy of litigation. There's certainty. The government says, if you do X, Y and Z, you get to drill. Here you do X, Y and Z, and then some appellate court decides there's a discrepancy and you are out of business."

By February 2011 world oil prices were rising steeply. Libyan oil fields had shut down. The US debt was rising. Congress could not even agree on a budget, let alone on energy policy. Perhaps the instability in the Mideast—perhaps the whole post–Wall Street meltdown and edgy economic situation in the world—would help bring exploratory drilling to the US Arctic, he hoped.

After a while Pete's depression lifted.

It took a few weeks, but he was back to normal.

"I'm optimistic for 2012," Pete said. "I am!"

Edward intended to make his last eight months in office count. With *kivgiq* over he plunged back into the oil fray, except now he was actively trying to pave the way for some offshore exploration in 2012.

At the Alaska Eskimo Whaling Commission convention that February he urged the captains to stay out of court when it came to challenging Shell.

"Ask yourself who is behind the decision to say no all the time? Is it Iñupiaqs sitting down together and hashing it out and taking control of our position? Or is it people from the outside who may have a very different agenda? I'm talking about the environmental groups that are in complete control of these lawsuits...Maybe we need to listen to each other for a while and politely ask the lawyers to leave the room."

He was also furious over the federal designation of critical polar bear habitat on the North Slope, an issue over which, while Shell and Itta were not actively aligned, the company quietly rooted for Itta. With the critical habitat stretching across the borough's coast, even if the EPA gave Shell a clean-air permit in 2012, the new rules would be used by opponents to try to stop Shell. Itta hated the designation for several reasons. It would block any development in too many places, he felt.

"I've joined in giving a notification of intent to litigate," Itta said.

"We were never consulted on this. We are not part of the bear recovery plan. I found out two hours before the formal notice that critical habitat was going to be listed, and when I got the announcement I about fell off my chair at 188,000 square miles of our region. These federal agencies need to get the message. We need to be part of the discussion."

He added, "Now Fish and Wildlife is looking for a spot in the federal register for a decision on the walrus. There will be a discussion on the bearded seal, whose skin we cover our boats with. And the ringed seal, which we eat.

"It seems like a tidal wave," Edward said. "The federal government dealt with the Arctic at first by ignoring it and then dealing with it in a slipshod, haphazard, totally confused method. Our government is behind the curve."

By now he was talking about more than oil or animals in the Arctic but as a citizen of the United States who was proud, and "got a high," he said, from being an American.

"There's a lack of infrastructure and preparedness. The Arctic seems to be on everybody's mind, but there doesn't seem to be any action. For the security of the nation I'd think this would be at the top of the list, I mean, oil and gas are two distinct items but there are also issues of marine transportation, issues of security. The tourist ships that came last summer? We mentioned that to the Coast Guard and they said, *Really? When was this?* I told them all these people just came ashore and were wandering around.

"Ours is the challenge to protect the land and the water that has always sustained us at the same time that we adapt to new uses," Edward told the Borough Assembly in his State of the Borough Address the following week.

The speech was delivered in English and the Iñupiaq language, but there were some phrases in it that had no corresponding meaning in Iñupiaq, such as "ten billion barrels."

Edward's wish as he began his last eight months in office was for his people and for—in a way—his legacy.

"I believe future generations will thank us as we thank our past leaders for having the foresight and wisdom to make the decisions they did way back when. To get us to where we are today," he said.

Edward promised, "I will continue to pursue this."

Even as his term was ending, a whole new round of Arctic issues faced the US and the North Slope. On Presidents' Day 2011, the men from the Department of the Interior arrived in town. They set up a chalkboard in the heritage center. They waited for anyone who wanted to participate in one more public comment period to walk in.

It was a cold night, clear, and fresh snow coated the streets outside; the rooftop of the library next door, built by taxes on oil companies; and the old folks' residence, built by oil money. And after a while a few cars and four-wheel-drive trucks parked outside, and even a bicycle. Barrowites were sick of these public comment periods, but this one was to precede a new Department of the Interior plan to block or allow drilling off the North Slope between 2012 and 2017. A whole new plan for as-yet-unleased areas.

The old Beaufort Sea leases, still undrilled, had been issued to Shell under DOI's 2002–2007 plan. The Chukchi leases, still blocked from being drilled, had been sold under the department's 2007–2012 plan.

With those fights still pending, now it was time for DOI to come up with the *next* five-year plan for the whole US continental shelf.

The men from the Department of the Interior would spend a few days flying to villages seeking comments, and then would go to Anchorage to seek more. Some of them had joined DOI after the *Deepwater Horizon* explosion. Most had worked for MMS before it became BOEMRE. One man had even been flown up from New Orleans to sit in, although that office never dealt with ice or bowhead whales. The Department of the Interior had spent many thousands of taxpayer dollars on this trip—for the stenographer present and hotel rooms and food bills and air tickets, for all the meetings in all the villages and all the uncoordinated deliberations that were well intentioned, mandated by law, but which after years of babble often seemed an end in themselves. No one from other agencies—EPA, NOAA, Coast Guard—was present.

Barrow's heritage center is a museum unlike most in the sense that unlike, say, Chicago's Field Museum, it does not bring in works from outside to show local people. It exists to show off the history of local people to visitors from outside.

There were exhibits on whale hunting, and photos of whalers, including Edward. There were traditional Eskimo masks and weapons and harpoons and boats. There was even a rank fishy odor permeating the center that night because in a back room workshop whaling crews were constructing an *umiaq*. The smell was sealskin.

After a while it became clear that most of the chairs in the room would stay empty—maybe because it was a national holiday and federal officials never held meetings on those days in other places, maybe out of long-term public meeting fatigue and disgust—so the men from the Department of the Interior, almost outnumbering the audience, began.

Taking his place in front was Jeffery Loman, a tall, casually dressed man who told the audience that the new five-year plan might in the end allow new Arctic drilling or might not. "It's a clean slate. Unlike the past, where previous administrations proposed specific numbers of oil or gas lease sales in the Chukchi or the Beaufort, we have no preconceived amount of oil and gas lease sales in either of those planning areas."

All options were on the table and tonight's effort, he said, was part of a national one in which BOEMRE staffers in other states also solicited comments to be considered in coming up with the 2012–2017 plan.

He mentioned Shell then. I'd heard him mention Shell in another talk in town too. He said that when Shell *decided* not to drill in the Arctic in 2011—as if the company had reached that decision at the exact moment when the way was clear—2011 became the ninth year in a row that no drilling occurred off the North Slope.

Were a stranger to have walked into the room at that moment, he or she would have thought that the debate was just beginning, that the government men were in Barrow for the first time, that the issue of offshore oil was in its infancy and that the night's event celebrated great promise and movement, great balance and hope.

It was easy to get that misimpression. All they had to do was listen to the man in front.

"What may be larger than the largest oil field on earth is believed to be out there," Jeffery Loman said.

Conclusion: The Real Enemy

When Edward Itta became mayor of the North Slope in 2005, he opposed all offshore oil extraction in the US Arctic. By 2011 he believed that some exploration was necessary if his people were to enjoy the same amenities enjoyed in other US communities— decent schools, emergency medical care, roads, heated homes, basic plumbing.

Itta had changed positions only after he helped force one of the world's largest oil companies to modify its original Alaska plans, which he had deemed too dangerous to his people. Along with groups including the Alaska Eskimo Whaling Commission, the Village of Point Hope, REDOIL and several national environmental organizations, Itta blocked Shell in court. He fought for the Department of the Interior to lay down new precautionary standards for offshore drilling. He negotiated a science agreement with Shell to fund research and provide baseline information about Arctic wildlife, climate and ice conditions, which will help North Slope mayors, the state of Alaska, oil companies and the US government plan for change in the High North.

275

Itta, in short, changed his mind *after* he helped force Shell to its knees.

Pete Slaiby had come to Alaska in 2008 hearing that North Slope residents in general and Edward Itta in particular were difficult, impossible to please. By 2010—still confident in his company—he had also come to respect the reasons why Shell's initial plans had faltered. He helped modify those plans along lines suggested by Itta and other Iñupiat groups. He signed the science agreement with Itta.

As a result of movement on both sides, by 2011 Shell's two most powerful Eskimo opponents in the US—the Borough of North Slope and the Alaska Eskimo Whaling Commission (with the latter's position contingent on a clean-air permit's going through)—dropped courtroom opposition to limited exploration in the Beaufort Sea. They would have preferred opening the potentially oil rich ANWR reserve to drilling but as that was blocked, they were looking north, believing that some offshore drilling was inevitable.

The Eskimo and the Oil Man had met in the middle, morphing from antagonists into unofficial, sometimes uneasy allies.

Their mutual enemy was now the US federal system for offering up oil leases and processing drilling applications offshore, which had evolved piecemeal under both Democratic and Republican White Houses and Congresses. The system—originally conceived of as one of checks and balances—had by 2011 degenerated into nonstop checks halting progress. The Byzantine labyrinth of uncoordinated laws, agency rules and court decisions, taken together, institutionalized the blocking of headway and made the lack of clear policy the butt of jokes at best and a springboard for never-ending fights.

Now Washington ought to follow the example of the Eskimo and the Oil Man. Unless ultraconservatives who clamor for offshore drilling at any cost and ultraenvironmentalists opposing it in all forms can reach the same point as Edward Itta and Pete Slaiby—*compromise*—America will never learn whether trillions of dollars' worth of recoverable oil and gas actually lie under Alaska's continental shelf, enough energy to fuel the nation's economy and bolster domestic supply until viable future alternative power forms come into existence.

The "drill, baby, drill" crowd needs to tone down the rapacious rhetoric and blind demand for universal extraction. The rampant Greens need to stop pretending that they represent broad Alaskan native interests, as they've historically advocated limiting the native relationship with the wild when it comes to whale hunting, bird hunting, polar bear habitat, land use.

Historian Shelby Foote once said that true American genius lies in the ability of its people to compromise. When it comes to oil exploration in the US Arctic, that genius is needed now.

William Reilly, former co-chairman of the Deepwater Horizon Commission, was still thinking about the Arctic in July 2011, seven months after the commission released its report calling for changes in the system by which the federal government leases and regulates offshore oil extraction.

Reilly explained over the phone that back in the winter of 2010 he recommended to President Obama during a White House meeting that Shell be permitted to drill a single offshore exploratory well in the Beaufort Sea in summer of 2011.

He was not supporting *extraction*, just, at that moment, exploration.

He said he told the president that when it comes to oil-spill cleanup, "Shell has the best preparation and response plan that anyone has ever put together. They'd have a tanker standing by in the event of a spill. Nobody has done that before."

He added, "It would take very little time to drill a relief well if you needed to do that...The whole project would only take 28 days and be finished by the time the ice moved in...There's a difference between production drilling and exploratory drilling."

By summer 2011 Reilly had spent months talking to oil company officials and critics, and he said that he'd also told President Obama, "The oil industry has not yet developed the technology adequate for cold-season production drilling in full ice. But they won't develop it unless they know that it is worth it. That's why we need to go forward with the minimal risk of purely exploratory drilling, a modest plan. We need to determine what the resource is, what is there."

Reilly called for more scientific research in the Arctic but no drill moratorium while it proceeded.

As a former head of the World Wildlife Fund and also of the EPA under President George Herbert Walker Bush, Reilly had plenty of experience with Washington infighting. He'd had numerous run-ins with Bush's chief of staff, John Sununu, over environmental policy during that administration. Reilly believed there was a human influence in global warming—that emissions from cars and factories contributed to it—and he pushed President Bush to cut US carbon emissions. Sununu vehemently opposed him.

By July 2011, Reilly's Washington experience made him suspicious of the EPA's record of blocking Shell's clean-air permit applications in Alaska.

"Given what Shell paid for those leases, they've been jerked around," he said. "Their ten-year lease is, what, five years old now? That doesn't seem justifiable as public policy. It is unfortunate that the EPA is allowing itself to be used as an instrument blocking development. There's something," he said, intimating that the holdups were politically motivated, "going on up there."

Reilly called Shell's 2011 drill plan "the gold standard" and said that he'd recommended to Secretary Salazar that the Department of the Interior "simply write into regulations for all companies the ambitious, responsible plan that Shell has."

As the conversation progressed it became impossible for me not to consider my own status as a reporter fly on the wall—in both camps—over the previous year. I told Reilly that in researching this book I had arrived at a conundrum. As a writer who regularly covered nature and climate, and as a believer in global warming theory I'd started the project intending to be objective but also suspecting that once the research was done I'd come away believing that the Eskimo and the Oil Man would remain antagonists and that my conclusions—for whatever an independent observer's are worth in a polarized country—would put me in the environmentalist camp.

What happened, however, was different and had caused me to agree with Reilly.

Reilly was amused at this. "Going to make lots of friends when the book comes out, huh?" he said.

America needs to think of itself as an Arctic nation. Only then will it take advantage of opportunities and prepare for the consequences of ice melt in the north.

To do this on the energy end, the US must streamline the process by which oil leases are granted and exploration licensed offshore. *If* recoverable oil or gas is really found, a whole new set of precautionary rules will kick in before extraction is permitted. Wouldn't it be wise for the nation to find out what is really there?

To accomplish this goal, Congress should consider overhauling environmental laws dealing with offshore drilling into a single comprehensive package that combines all rules into one framework and provides guidance, one-stop oversight and adequate funding for regulators. Congress ought to also allow the limited testing of oil cleanup methods in Arctic waters to see whether they really work.

The overhaul must occur without creating new federal bureau-cracies. Existing agencies must be required to make decisions together, in a timely manner, so oil companies that buy leases get clear instructions on what preparations are necessary before they drill. Either that, or don't sell leases in the first place. Permanent advisers to decision makers should include representatives from appropriate agencies, the oil industry, Alaskan native groups and national environmental ones.

As in Norway, permanent appointees who do not change with administrations should head the separate government oversight groups now responsible for leasing and for safety. That would cut down on politics and keep policy more stable while promoting environmental protection, regardless of which political party occu-pied the White House at any given time.

If recoverable oil is located offshore, taxes should be raised on profits to help fuel America's recovery and pay for research on alternative fuels. While it is true that Shell sank $3.5 billion into Alaskan offshore leases without getting to drill between 2007 and 2010, it is equally true that this huge sum represented just half of Shell's fourth-quarter profits in 2010. Shell's reported profits world-wide in 2010 were $18.6 billion, up from $9.8 billion in 2009.

Instead of wasting hundreds of millions of dollars on do-over plans, endless uncoordinated public hearings, lawyers and lobby-ists and renting equipment that never gets used, wouldn't it be bet-ter for these monies to go toward public works, energy research, education and jobs throughout the US?

Oil companies can and should pay more in taxes once oil is found.

The choice that the US empire faces—use the Arctic or lose it, as Canadian prime minister Stephen Harper has said—is not unlike the one made by President Andrew Johnson's secretary of state in

1867, when William Seward purchased Alaska from Russia for $7.2 million. Most Americans thought the deal worthless and called it Seward's Folly, or Seward's Icebox.

After all, there were more immediate needs to deal with, they felt. The Civil War had just ended and the US was saddled with war debt and reconstruction costs. Washington owed money for hundreds of thousands of military pensions. Just the bill for the fighting topped $15 billion.

But Seward foresaw the economic and political importance of the north. Later he was proven right.

With the discovery of gold in Alaska and, later, oil, Alaska contributed hugely to the American economy. It also played a key role in national security. The Aleutian Islands provided a buffer against the Axis powers during World War II. The DEW Line radar system guarded North America during the Cold War and still does today.

"Seward got a lot of criticism for buying Alaska, but a hundred years later that land that was bought for two cents an acre ended up being the largest provider of energy in the US," Mead Treadwell told me. "Now in the next few years we're going to spend $60 million of taxpayer money to make a claim for land in the Arctic Ocean that may be greater than the size of California. We may not see the benefit of that in our lifetime, but if Seward hadn't purchased Alaska, imagine what the world would have been like. Imagine the Cold War if Russia had owned Alaska. America may not understand the value of making our claim in the Arctic Ocean in 2011, but we will get a good return for it."

To underline the importance of the Arctic, Treadwell recalled an incident that took place when he was Alaska's deputy commissioner for the environment in 1994, during a joint federal and state emergency simulation drill held to practice government response in the event of a major earthquake.

As part of the simulation, Mead said, "I shut down the Alaska pipeline so we could inspect it and make sure there weren't any leaks. Within *ten minutes* I heard from the secretary of energy in Washington. He reminded me how strategic the pipeline is and why we need to keep it open."

In other words, Washington was concerned that even a brief shutdown would damage the nation. Yet in 2012 the same pipeline is threatened with total closure from lack of product—not from an earthquake or a terrorist attack but from political gridlock preventing new supply from flowing.

The practical effect of trying to skirt national environmental laws—as occurred in agencies when George W. Bush was president, or of trying to halt offshore drilling as happened under Barack Obama—has had the same effect in the end.

No oil flows.

Today Seward's Icebox is melting and so is America's strength as the nation allows a historic opportunity to slip away and as politicians spend time arguing about *why* warming is occurring instead of dealing with the fact.

As Mead Treadwell said, "There's no climate theory that brings the ice back."

America needs to embrace itself as an Arctic nation. It is the only nation on earth that has successfully incorporated far-distant lands and peoples into the national fabric. Past empires owned colonies overseas but these areas—unlike Alaska and Hawaii—were not full-fledged parts of the nation. Their citizens did not enjoy the same rights as everyone else.

America has done a splendid job of joining distant places to the main part of the country. But it has done a poor job of grasping the challenges and benefits of the arrangement when it comes to the High North.

"The Arctic has been out of sight, out of mind," Scott Borgerson said at the Council on Foreign Relations in New York. "It could be the dark side of the moon to most Americans. But the issues on the North Slope will affect every American."

These are not future challenges but current ones. "There are decisions we need to make in the Arctic today," Mead Treadwell said. "Some people say, I don't want anything new to happen in the Arctic. But that's putting your head in the sand. All you change is your own ability to play later."

"It is magnificent to stand on the sea ice, magnificent to stand in a valley and be the only human around," he added. "To see the polar bear in the wild. Foxes. Bird colonies. The call of the north is still there. But at the same time we better not deceive ourselves, because this is an area vital to US commerce and the strength of our country. There are very few places left on earth where you can say, if you do the right things now you can also do the right things later, sustainably, but the Arctic is a case in point."

To prepare for an opening Arctic the US must ratify the Law of the Sea Treaty in order to have a say in how the seabed around the world is divided up, and in order to submit American claims and participate in the way the world governs the oceans. In 2012, Russia is expected to press its claims for annexation of 380,000 square miles of Arctic. Russia plans to build six more icebreakers and spend $33 billion to construct a year round port on its Arctic shores. Russia and Norway have signed a historic agreement delineating areas they control in the Barents Sea. All while the US dithers.

Alaska's politicians understand the stakes. While it would be easy to dismiss their efforts as mere attempts to send some DC pork home, their initiatives have lots of merit.

Republican senator Lisa Murkowski requested money in 2009 for a study by the Army Corps of Engineers of locations for a possible US Arctic port. She also introduced a bill to improve navigation abilities in the High North.

Alaska's Democratic senator, Mark Begich, introduced a package of bills to encourage adaptation to changes in the region. His Arctic Marine Shipping Assessment Bill, passed by Congress in the autumn of 2011, will promote more research in the High North. He also sponsored legislation requiring the secretary of commerce to research ways to prevent or respond to oil spills offshore.

Since all parties agree that more science is needed in the Arctic, NOAA should receive more funding for research in order to gather crucial baseline information about high-latitude ecosystems. But these studies should be conducted *while* some offshore hydrocarbon exploration is proceeding. Studies begun now will provide information in time to affect eventual plans—either the allowing or the blocking—of production wells in the offshore Arctic.

Some of these new studies could be conducted jointly by oil company, government agency and North Slope researchers. A boilerplate project is the science research agreement signed by Pete Slaiby and Edward Itta in 2010. The resulting work will not be owned by either the oil company or the North Slope Borough and will therefore reach the public more quickly. The more that is known about the Arctic, the smarter the decisions that will be made there. Joint research proves that goodwill exists among parties, demonstrates a willingness to be sensitive to environmental issues and pushes along the best possible oil-exploration process.

Another excellent example of cooperation—in this case between industry and government—occurred in summer 2011 when Shell, ConocoPhillips, and Statoil signed an agreement with NOAA under which all parties will share ocean, meteorological and coastal research data in the US Arctic.

Finally, when it comes to government initiatives, the Coast Guard and Navy need to purchase icebreaking ships and communication systems for the High North, so as to be present if offshore oil exploration begins, and as new shipping, tourism and security issues confront the region. All service branches need improved ability to operate in the Arctic.

The rapidity with which the High North is presenting Arctic countries with security challenges was underscored in September 2011 during two days of war games held at the US Naval War College in Newport, Rhode Island, the third time in a fifteen-month period that strategy exercises anticipated operations in the north.

Outside, the temperature reached 75 degrees those days, while inside, in the college-style setting, in classrooms and a larger meeting room, naval planners, academics, ice scientists and technical experts—divided into groups—considered future scenarios posited by game planners in a world with *less ice*. In one, a ship carrying terrorists was moving through the Northwest Passage. It needed to be boarded. In another, a merchant vessel out of North Korea carrying weapons of mass destruction sank in the Chukchi Sea, northwest of Barrow. The weapons needed to be recovered. A third scenario involved a massive oil spill erupting north of Barrow, coming from an offshore well. The spill needed to be cleaned up and the Navy needed to assist.

The specific scenarios in the games were less important as actually anticipated events than as a means to get participants "thinking about moving things around in the Arctic box," game designers said.

Starting at 8 a.m. each morning, men and women in several large meeting rooms—working at computer screens—tried to figure out proper questions to plan for when anticipating Arctic missions. How to use suitable ports; how to refuel in Arctic conditions; how to clothe, feed and lodge hundreds of sailors or troops or cleanup workers in Arctic conditions. How to coordinate US and Canadian military action inside the Northwest Passage at the same time that Canada claims the passage as internal waters and the United States regards it as open to all.

Where were the gaps? Where did the Navy need more intelligence, equipment, knowledge? How could ships move north without disrupting marine mammals or interrupting Iñupiat hunting off the North Slope, if possible?

During the oil-cleanup game, I watched over Cmdr. Kris Moorhead's shoulder as Google Earth suddenly showed Barrow. Moorhead eyed the roads, rooftops and shoreline of the city. On numerous screens that day the eyes of the US Navy scrutinized the AC Value Center and ASRC building and BASC and flicked over Mayor Itta's house and the Utilidor pipe, as well as the very route that Itta had driven his snowmobile over on the way to meet Pete Slaiby and President Marvin Odum of Shell Oil in May of 2010 during the *Deepwater Horizon* disaster.

Itta had envisioned an oil spill then. On a late summer day a year and a half later, planners in Newport did too.

Barrow slept, four hours behind Newport. Nine a.m. East Coast time was 5 a.m. on the North Slope. As Itta dozed, planners from the most powerful Navy on earth tried to figure out if Arctic ice would hinder ships working on oil cleanup in the Chukchi Sea, if a spill lasted into autumn. If they should recommend using explosives against ice. And how the United States could obtain icebreaking ships to lead the way during Arctic military missions.

No one had thought to include Eskimos in the exercise. During several sessions, when questions about Eskimos came up, game participants consulted Canadian Coast Guard people for answers, as if Canadian and US Eskimos had the same views on everything and they could be articulated by non-Eskimos in the room. Finally an outside observer suggested to game planners, in private, that next time an Arctic game was held it might be a good idea to invite people who live in the High North to participate.

The response was, "I wish someone would have suggested that three months ago."

Meanwhile, as questions multiplied, they were typed up to be considered by a larger plenary group, and eventually evolve into recommendations for the Navy. How many icebreakers would be required to lead the way if a fleet needed to transit the Northwest Passage, going from San Diego to Virginia, as they were doing in scenario number one? Should the US rent icebreakers from other countries or build them or simply strengthen existing ships?

Should the US ask Russia or Canada for help in an emergency, and by what procedure?

How would ocean currents move radioactivity through the region if a nuclear device contaminated them, as in scenario three?

How would the United States coordinate a response to a terrorist attack or oil spill with Denmark, Norway or Russia?

These were not fantasy challenges but actual *anticipated problems* of the near future in America's High North.

"Right now the Navy doesn't have a mission in the Arctic. But it is coming," said Capt. Kevin Hall of US Fleet Forces Command.

Empires rise by taking risks and having clear purpose. They decline through inaction, infighting and gridlock. Former US secretary of state Henry Kissinger said, "When one is on a tightrope the most dangerous course is to stop."

Yet true progress has stopped in the US Arctic. A distracted America is handing away control of the region. The effect will extend outward to the nation and world.

Alaska state legislators were prescient in 1958, when statehood came along and they sent a message to Washington by designating the state flower as the forget-me-not.

Now Washington squabbles over energy possibilities in the High North while US rivals grow rich on oil profits. Iran, Russia,

Colombia and Saudi Arabia all fund their international influence with US petrodollars.

As I write this, America is in the throes of a presidential election year. Energy policy will be a huge part of the campaign. The Arctic offers the nation fresh ways to tackle energy needs and financial woes and ways of reasserting dominance in the world. It poses a challenge when it comes to protecting a vast, pristine environment and native peoples who live there. It presents itself as a proving ground for American ingenuity and vision.

The Arctic in the 21st century is a place defined by the words of Robert Kennedy as he contemplated the nation's role in international affairs decades ago.

"Great change dominates the world, and unless we move with change we will become its victim," he said.

As the Eskimo and the Oil Man know, that seems true today in the Arctic as much as anywhere on earth.

Acknowledgments

The kindness of many people allowed me to complete the research and the writing of this book. I would like to thank them here.

Thanks to Mayor Edward Itta and to Shell Oil's Pete Slaiby. Both shared their stories, thoughts and access to their staffs with a stranger. Thanks for the trust.

In Barrow, a special thanks to Richard Glenn at ASRC, to Geoff Carroll, and to Glenn Sheehan, Brian Thomas, Lewis Brower and Nok Acker at the Barrow Arctic Science Consortium. At the North Slope Borough Department of Wildlife Management, Taqulik Hepa, Robert Suydam, Jason Herreman, Craig George and Michael Pederson went out of their way to be helpful.

In Mayor Itta's office, a very special thanks to David Harding, Carla Kolash, Andy Mack, Marie Itta and Harold Curran.

At Sandia National Laboratories, a thousand thanks to Valerie Sparks and to Bernie Zak for their kindness.

At Shell Oil's Alaska Venture, thanks to Curtis Smith, Susan Childs, Michael Macrander, Geoff Merrill, Tom Homza, Cam Toohey, Crystal Mason and Michelle Romak.

In Anchorage, I cannot thank Alaska's lieutenant governor, Mead Treadwell, enough.

Thanks to the US Coast Guard for providing access to the icebreaker *Healy* during two trips, and also for allowing me to come along on Operation Arctic Crossroads. Thanks also to former commandant Thad Allen, and admirals Gene Brooks and Chris Colvin, as well as Cmdr. Michelle Webber. On the *Healy* and also in Seattle, thanks to my *Healy* cabinmate Marcus Lippmann, who made me feel welcome in both places.

I was fortunate when I got interested in Arctic matters to work with several terrific magazine editors who sent me north over a three-year period. Enormous thanks to Daryl Chen at *Parade* magazine, to Janice Kaplan, also at *Parade*, to Carey Winfrey at *Smithsonian* magazine and to Chris Keyes at *Outside* magazine.

A huge thanks to my editor John Brodie at Grand Central Publishing, who saw the value of the idea for this book and who helped guide the research and writing as the story twisted and turned.

I'm lucky to have good friends like Ted Conover and Phil Gerard, and to have cousin Meryl Cohen, all of whom read drafts and pointed out places where the work could be better.

In Washington, DC, thanks to Jeff Stein, Jim Grady and Bonnie Goldstein for letting me stay with them when I was in town.

In Rhode Island, thanks to Bob and Kathy Leuci for providing a warm and welcoming house.

For the eighteenth time—I'll never get tired of this—thanks to my great agent, Esther Newberg.

Jana Goldman at NOAA is one of the best public relations specialists I've ever had the pleasure of working with. Thanks also to Robert Freeman at the US Navy and the whole staff at the Coast Guard.

For anyone interested in US Arctic history and related issues, I'd like to recommend several terrific books that helped in the research. They are *Fifty Years Below Zero*, by Charles D. Brower; *The Firecracker Boys*, by Dan O'Neill; *The Whale and the Supercomputer*, by Charles Wohlfort; *Fifty Miles from Tomorrow*, by William Hensley; *Resolute*, by Martin Sandler and *Watching Ice and Weather Our Way*, by Conrad Oozeva, Chester Noongwook, George Noongwook, Christina Alowa and Igor Krupnik.

Any mistakes in the text are mine.

Finally, when it comes to key energy sources in the Western Hemisphere, none surpasses Wendy Roth, whose loving support and constant supply of doughnuts kept me going throughout the project.

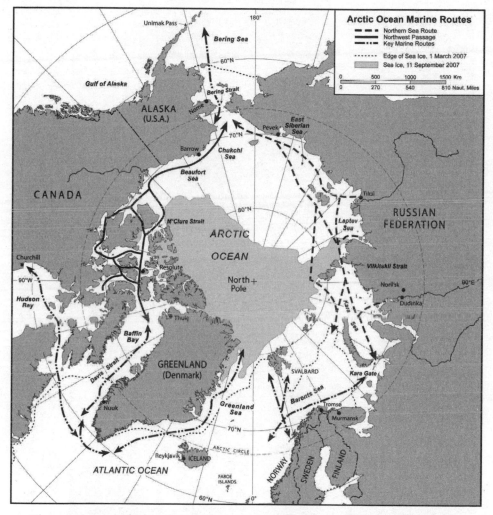

The melting of the Arctic ice cap is opening up new shipping routes and greater access to new oil and gas fields at the top of the world. Courtesy of Dr. Lawson W. Brigham, University of Alaska–Fairbanks.

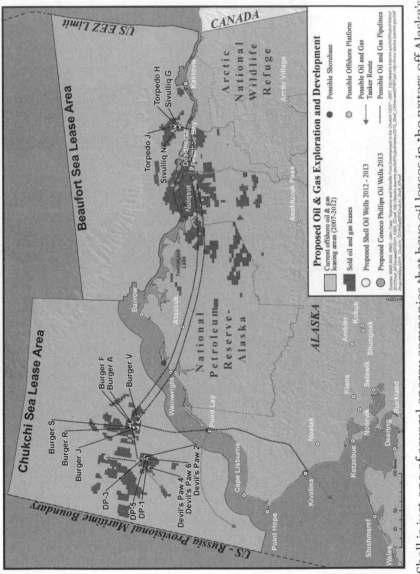

Shell is just one of several energy companies that have oil leases in the waters off Alaska's coast. Courtesy of Northern Alaska Environmental Center, 9/1/2011.

INDEX

About the Author

Bob Reiss is the bestselling author of 18 books of nonfiction and fiction and is published in twelve countries. He is a former *Chicago Tribune* reporter and former correspondent for *Outside* magazine. He's covered the US Arctic for *Smithsonian*, *Parade* and *Outside* magazines, and for *Politics Daily*. Bob lives in New York City with television producer Wendy Roth. You can learn more about Bob's books or public appearances at www.bobreiss.com.

**BUSINESS
PLUS**

Recognized as one of the world's most prestigious business imprints, Business Plus specializes in publishing books that are on the cutting edge. Like you, to be successful we always strive to be ahead of the curve.

Business Plus titles encompass a wide range of books and interests—including important business management works, state-of-the-art personal financial advice, noteworthy narrative accounts, the latest in sales and marketing advice, individualized career guidance, and autobiographies of the key business leaders of our time.

Our philosophy is that business is truly global in every way, and that today's business reader is looking for books that are both entertaining and educational. To find out more about what we're publishing, please check out the Business Plus blog at:

www.bizplusbooks.com